SpringerBriefs in Petroleum Geoscience & Engineering

The SpringerBriefs series in Petroleum Geoscience & Engineering promotes and expedites the dissemination of substantive new research results, state-of-the-art subject reviews and tutorial overviews in the field of petroleum exploration, petroleum engineering and production technology. The subject focus is on upstream exploration and production, subsurface geoscience and engineering. These concise summaries (50-125 pages) will include cutting-edge research, analytical methods, advanced modelling techniques and practical applications. Coverage will extend to all theoretical and applied aspects of the field, including traditional drilling, shale-gas fracking, deepwater sedimentology, seismic exploration, pore-flow modelling and petroleum economics. Topics include but are not limited to:

- Petroleum Geology & Geophysics
- Exploration: Conventional and Unconventional
- Seismic Interpretation
- Formation Evaluation (well logging)
- Drilling and Completion
- Hydraulic Fracturing
- Geomechanics
- Reservoir Simulation and Modelling
- Flow in Porous Media: from nano- to field-scale
- Reservoir Engineering
- Production Engineering
- Well Engineering; Design, Decommissioning and Abandonment
- Petroleum Systems; Instrumentation and Control
- Flow Assurance, Mineral Scale & Hydrates
- Reservoir and Well Intervention
- Reservoir Stimulation
- Oilfield Chemistry
- Risk and Uncertainty
- Petroleum Economics and Energy Policy

Contributions to the series can be made by submitting a proposal to the responsible Springer contact, Charlotte Cross at charlotte.cross@springer.com or the Academic Series Editor, Prof. Dorrik Stow at dorrik.stow@pet.hw.ac.uk.

More information about this series at http://www.springer.com/series/15391

Shivanjali Sharma · Amit Saxena ·
Neha Saxena

Unconventional Resources in India: The Way Ahead

Springer

Shivanjali Sharma
Department of Petroleum Engineering
Rajiv Gandhi Institute of Petroleum Technology
Jais, Uttar Pradesh, India

Amit Saxena
Department of Petroleum Engineering
Rajiv Gandhi Institute of Petroleum Technology
Jais, Uttar Pradesh, India

Neha Saxena
Department of Petroleum Engineering
Indian Institute of Technology (ISM)
Dhanbad, Jharkhand, India

ISSN 2509-3126 ISSN 2509-3134 (electronic)
SpringerBriefs in Petroleum Geoscience & Engineering
ISBN 978-3-030-21413-5 ISBN 978-3-030-21414-2 (eBook)
https://doi.org/10.1007/978-3-030-21414-2

This Springer imprint is published by the registered company Springer Nature Switzerland AG
The registered company address is: Gewerbestrasse 11, 6330 Cham, Switzerland

Contents

Part I
Coal Bed Methane (CBM)

Prospective Evaluation and Prediction of Coalbed Methane Production from Fields of Indian Sub-Continent

Background

Coalbed methane (CBM) has gained considerable ground as an unconventional source of energy in the recent past. CBM which was considered to be uneconomical and nonconventional only a few years back has now become very much attractive as a new energy resource. Specifically, for the countries like India where more than 75% of the total energy demand is accomplished from imported oil and gas, CBM can play the most important role to sustain its rapid pace of development.

Coalbed methane (CBM) is natural gas, and it is generated during coalification process and get adsorbed in coal at high pressure. CBM is rich in methane (88–98%) which emanates from coal due to change of in situ pressure conditions. Methane which is found in coal seams is named as coalbed methane. Large amounts of gases like methane, ethane, CO_2, water vapor, H_2S are produced during coalification, and a portion of them is held both in the coal seams and adjacent rocks (Kim and Kissel 1988; Patching 1970).

Methane is the principal gas in this mixture. Methane which is 23 times more potent greenhouse gas (GHG) than CO_2 leads to mining hazard if not ventilated prior to coal mining operations. Thus, the presence of CBM in underground mine not only makes mining works difficult and risky, but also makes it costly. On the other hand, CBM is a remarkably clean fuel when burnt with a heating value of approximately 8500 Kcal/Kg compared to 9000 Kcal/Kg of natural gas of power grid quality. Thus, CBM production not only can provide additional energy to fulfill more demand, but also help to reduce global warming to great extent. Enhanced recovery CBM by CO_2 sequestration can improve the situation.

Worldwide total coalbed methane potential has been estimated to be 89 trillion m^3 to 269 trillion m^3 of gas in place (Charles et al. 1998). Countries like USA, Australia, China, and India are currently producing CBM on an economic scale. USA is the largest CBM producer in the world, and 1.91 TCF gas have been sold in 2009 (Pashin 2011). India holds the fourth largest proven coal reserves (Coal Atlas of India 1993) and third largest coal producer country in the world. Indian coalfields are divided into two broad groups of two distinct geological ages: Gondwana coalfields of Permian age and Tertiary coalfields of Tertiary age. More than 98% of Indian coal production comes from Gondwana coals. These coalfields belong to the Damodar Valley (West Bengal—Jharkhand), Son—Narmada valley (Madhya Pradesh), Mahanadi Valley (Orissa), and Pranhita—Godavari valley (Maharashtra–Andhra Pradesh). Raniganj and Birbhum coalfield in West Bengal; Bokaro, Jharia, North Karanpura, and South Karanpura coalfield in Jharkhand; Singrauli, Sohagpur, and Satpura coalfield in Madhya Pradesh; Ib-valley and Talchir coalfield in Orissa; Korba in Chattishgarh and Wardha in Maharastra are the most important Gondwana coalfields having vast reserve of good quality coal. Tertiary coalfields are Assam–Meghalaya; Neyveli; Cambay; Barmer—Sanchor; Bikaner and Jammu and Kashmir. On the basis of coal rank, maturity, Physicochemical attributes of coal, depth of occurrence of coal, available area and CBM potential, Indian coals are divided into four types: category-I, category-II, category-III and category-IV (Fig. 1). Jharia, Bokaro, North Karanpura, and Raniganj coalfield belong to category-I type and comparable with global producing CBM fields in terms of gas content and adsorbed capacity (Hajra et al. 2003) for their good coal seam thickness, high rank, and maturity. Damodar and Mahanadi valley coalfields are categorized as category-II and category-III. The above-mentioned Tertiary coalfields are placed in category-IV as their CBM prospects are yet to be established.

India has started evaluating different coal-bearing sedimentary basins for their CBM potential in the early nineties (Patra et al. 1996). Exploration and development of coalbed methane in Jharia and Raniganj coalfield have been going on for about last 15 years and being actively explored by different exploration and production companies. India has a prognosticated CBM resource of around 92 (in TCF) of CBM reserve (DGH report 2016–17) which may fulfill the country's future growing energy demand to a large extent. However, detail information of Indian CBM field is lacking in the literature. Jharia coalfield of Jharkhand, Raniganj coalfield of West Bengal, and Singareni coalfield of Andhra Pradesh have been selected for evaluation of chemical parameters, gas content, gas adsorption capacity, gas saturation, and recovery of gas by primary and secondary processes. In situ gas content measurements have been performed for two coalfields, i.e., Jharia coalfield and Raniganj coalfield, from the wells under drilling. Most of the fields are still under exploration stage while a few have just started production.

The methane content of the Jharia and Raniganj has been estimated by direct gas content measurement or Canister desorption test. A more general estimate can be made using adsorption isotherm data. Adsorption isotherm curves indicate that gas adsorption increases with increasing rank of coal at a given temperature and pressure condition. Gas saturation and pressure at which gas can be started to release also determined from isotherm curves.

The initial recovery of CBM requires depressurizing through long-time dewatering and massive hydrofracturing of coal beds. Gas and water are produced

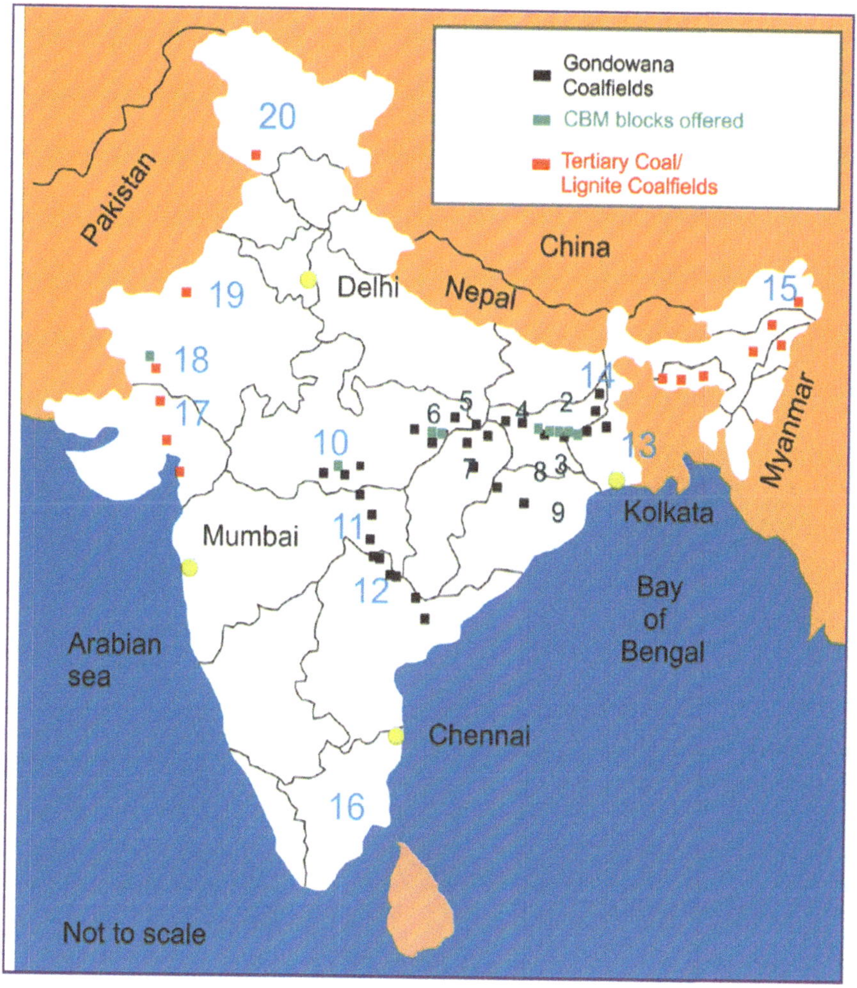

Fig. 1 Distribution of coalfields and CBM blocks, India

through cleats opening to the production well by lowering of pressure around the well, and initially, water production is more compared to gas production (Nuccio 2000). Gas starts to desorb by reducing the pressure at the matrix–cleat interface, gas diffusion occurs through the matrix to the cleat (Gunter 1997 and Law et al. 2003), and CBM is produced.

In India, the production of methane from Indian CBM fields still is in preliminary stage. Advanced research is required for efficient recovery from these fields. Few of the experimental works for enhanced coalbed methane recovery for Indian CBM fields have been presented by Prusty 2008; Dutta et al. 2011; Vishal et al. 2013; and Bhowmik and Dutta 2013.

Chapter 1
Introduction to Coal Bed Methane (CBM)

Abstract CBM is one of the most promising source of unconventional hydrocarbon. The following chapter is aimed to develop a primary understanding of coal bed methane. It will include its history, generation mechanism, coal assessment for CBM storage, production, composition of coal bed methane and finally its future in India.

Keywords CBM · Methane generation · Vitrinite reflectance · Petrography

1.1 The History of CBM

Coal Bed Methane is primarily the methane gas which is evolved during the formation of coal from the woody matter of plant origin. The gas is observed to get adsorbed on the surface of the coal. Production of coalbed methane (CBM) has become an important source of fuel providing a non-polluting fuel in an era focused on reducing the pollution. The shortages in the supply of fuel is also a matter of concern for many countries. Apart from methane CBM has variable amounts of carbon dioxide (CO_2), nitrogen (N_2), and other hydrocarbons, such as ethane (C_2H_6), and traces of propane (C_3H_8), and butanes (C_4H_{10}) (Gerald Tindal 2013; Scott et al. 1994).

1.2 Coalification and Generation of Methane

Methane is evolved in the coalification process. Coalification is defined as the formation of coal from woody matter of plant origin. The generation of CBM during coalification process is attributed to two principal ways as follows:

S. Sharma et al., *Unconventional Resources in India: The Way Ahead*,
SpringerBriefs in Petroleum Geoscience & Engineering,
https://doi.org/10.1007/978-3-030-21414-2_1

I. From biological process owing to the metabolic activities of different organism, and
II. From thermo-genic process i.e. resulting from the thermal cracking of the products having higher concentration of hydrogen (Thakur et al. 2014).

The methane is normally generated at a shallow depth of less than 10 m. Also the generating coal seams are thermally undeveloped as the temperatures are found to be less than 50 °C. Both the conditions of shallower depth and thermally immatures seams make favourable condition for biogenic CBM. It is done by the first process in low rank stage of coalification (sub bituminous) and is termed as biogenic or diagenetic methane (Thakur et al. 2014; Moore 2012).

Whereas when the methane being generated is under relatively low energy (pressure >10 m depth, and temperature >50–170 °C) conditions in the second step (catagenesis and metagenesis) is called as thermogenic methane. It is generated from the plant organic matter going under chemical or thermal degradation in the above mentioned window of temperature and pressure. Also, above this temperature the chances for generation of biogenic methane are minimal. This is attributed to the negligible activity of the bacteria at elevated temperature and pressures which doesn't support biochemical activity (Thakur et al. 2014; Moore 2012).

Most of the gases generated in the early biogenic stage escaped due to poor gas retention capacity of low rank coals and shallow depth of burial. Rank refers to the steps in the cooking of buried plant matter through natural process called "coalification," and getting transformed into a thicker, carbon rich, dry substance. The coal can be classified in four major ranks namely: Anthracite, Bituminous, Sub-bituminous and Lignite.

- Anthracite: Considered to be among the top most ranks of the coal. It is characterized by the following physical properties of hardness, brittleness and black lustrous structure. This is also known as hard coal which contains a higher percentage of carbon and a lower percentage of volatile matter (VM).
- Bituminous: This coal is considered among the intermediate ranks of coal after the top rank of anthracite. The coal is estimated to have a high heating value. This coal is among the most commonly used variety coal for electricity generation around the globe. Bituminous coal is characterized physically by a smooth and shiny structure. The closer looks make the layer structure of the coal prominent.
- Subbituminous: It's the third category of the coal which is characterized by a physically non-lustrous black coloured structure. The heating value of this coal is less than that of bituminous.
- Lignite: Considered to be the lowest rank of the coal. This is also known as brown coal having the least composition of carbon (Strąpoć et al. 2008).

Vitrinite reflectance is among the most widely used optical property for the classification of ranks of coal. It is defined as the percentage of incident light reflected from the surface of vitrinite particles in a sedimentary rock. Results are often presented as a mean R_o value based on all the **vitrinite** particles measured in an individual sample. As suggested by Scott et al. (1994), the vitrinite reflectance of the bituminous coal rank was found to be between 0.6 and 0.8%. From the experimental results it is evident that the CBM produced from the thermogenic process exhibits a maximum production at R_o ~ 1.2%. Also it was observed that the CBM production goes negligible when R_o ~ 3.0%. The reduction in the generation of CBM is attributed to the depletion of organic hydrogen pool. The rest of the hydrogen is largely found to be utilized in the aromatic structures. Therefore, it can be estimated that a larger amount of thermogenic methane was generated from higher rank coal as compared to that of the lower ranks. This will eventually end up having a higher CBM content if the gas remains contained in the structures. CBM from the Biogenic resources contributes to around 10% of the total methane. The rest of CBM that was produced during the coalification process is from thermogenic methods (Thakur et al. 2014).

Methane generated by biogenic process is dry while it is wet/dry when generated by thermogenic processes (catagenetic/metagenetic). When CBM is produced from the coal formations accompanied by water in its cleats, fractures and pore spaces the bearing formation is termed 'wet'. Whereas if the CBM bearing coal is free from water in its structure it is termed as dry. On the contrary the CBM produced from the thermogenic is because of chemical and thermal degradation above 100 °C. Above this temperature the microbial activity becomes negligible (Thakur et al. 2014). Thermogenic gas generation starts at the high-volatile bituminous coal. The vitrinite reflectance values were found in the range of 0.6 and 0.8% (Scott et al. 1994).

There is one more species of coal named "peat". It is not an actual coal, but often referred as the starting material for the formation of coal. It is a soft organic material that comprises of the dead plants along with some mineral content. When peat is subjected to elevated pressure and thermal conditions it instigate the formation of other ranks of coal.

Gases generated in the succeeding thermogenic stage could not migrate as a result of high-pressure regimes and remained stored in the coal. Generation of large amount of methane actually happens during the transformation of high volatile bituminous coal to low volatile bituminous coal stage. However, some portion of the generated methane is found to be retained in coal bed or seams. The gas is retained because of the prevailing high pressure. The excess of the generated gas migrates to the vicinity having probable holding capacity for the gas. It exists as monomolecular layer held up within the micro-pore structure of the coal. Most available coalbeds have in situ gas contents of 1–20 m^3/t (Scott et al. 1994; Thakur et al. 2014).

1.3 Composition of CBM Gas

Although methane (CH_4) is the major component of coal gases, other gases such as ethane (C_2H_6), propane (C_3H_8), butane (C_4H_{10}), carbon-di-oxide (CO_2), nitrogen (N_2), and water are released during coalification. Cumulative amount of methane that is formed during the coal formation (R_o max 0.5–1.8%) approximately ranges between 2000 and 5000 scf/t (Scott et al. 1994).

1.3.1 Coal Petrography

Coal petrography enables us to determine the relative abundance of different maceral and mineral matter in coal. Macerals are the smallest microscopically recognizable entities within the structure of coal. Coal consists of three main maceral groups: Vitrinite, Inertinite & Liptinite (Exinite). The above characterization can be sub divided into maceral sub groups and macerals. Minerals are the impurities in coal. The high Vitrinite content and very low Liptinite content of coal suggest that it is a gas–prone source rock. During coalification the generated coalbed gases are stored in macerals of coal mostly in adsorbed form with subordinate amount occurring as free gas within cleats and fractures and a small amount may be dissolved in water.

Vitrinite is the principle gas sorbet maceral in coal and have the largest surface area for Methane adsorption. Dominance of Vitrinite (>60%) results in high surface area in coals and thereby increasing Methane adsorption capacity. Vitrinite rich coals (bright coal) have greater Methane adsorption capacity than Inertinite rich coals (dull coals) of equivalent rank. This is due to Vitrinite having predominantly micro-pores (<20 Å or 2 nm) compared to Inertinite which supposedly has the dominance of meso- (>2–50 nm) and macro-pores (>50 nm). So coals with higher Vitrinite content are likely to have higher gas content, providing other parameters are also favourable. Thus coal petrography plays an important role in CBM exploration (Thakur et al. 2014; Scott et al. 1994).

1.3.2 Coal Rank

The rank of a coal indicates degree of coalification; the organic matter was subjected to. It is a particular stage in the maturation path of coal from peat to anthracite. Major rank parameters of coal include volatile matter content, fixed carbon content, moisture and Vitrinite Reflectance, these are known as rank parameters. Gas/Methane content of coal increases proportionally with the rank of a coal.

1.3.3 Grade of Coal

The grade of a coal refers to its degree of impurity. The major grade parameters affecting the sorption capacity of coal is mineral matter content. The dominant forms of mineral matter in coals are clay minerals, quartz, pyrite and calcite. Proximate and ultimate analyses of coals generate the key grade parameters (Levine 1993). These analyses report ash, which include the non-combustible residue of mineral matter and sulphur forms like pyretic sulphur and organic sulphur. Low grade coals have less adsorptive capacity for gases. Coal industry use grade parameters to decide utilization of a particular coal.

1.3.4 CBM Prospects in India and World

Worldwide total coalbed methane potential have been estimated to be 89–269 trillion m^3 of gas in place (Charles et al. 1998). Countries like USA, Australia, China and India are currently producing CBM on an economic scale. USA is among the largest CBM producer in the world (Pashin 2011). India among these holds the fourth largest proven coal reserves and third largest coal producer country in the world. Indian coalfields are divided into two broad groups of two distinct geological ages: Gondwana coalfields of Permian age and Tertiary coalfields of Tertiary age. More than 98% of Indian coal production comes from Gondwana coals.

India has started evaluating different coal-bearing sedimentary basins for their CBM potential in the early nineties. Exploration and development of coalbed methane in Jharia and Raniganj coalfield have been going on for about last 15 years and being actively explored by different exploration and production companies. India has a Prognosticated CBM Resource of around 92 (in TCF) of CBM reserve (DGH Report 2016–2017) which may fulfill the country's future growing energy demand to a large extent. However, detail information of Indian CBM field is lacking in literature. Jharia coalfield of Jharkhand, Raniganj coalfield of West Bengal and Singareni coalfield of Andhra Pradesh have been selected for evaluation of chemical parameters, gas content, gas adsorption capacity, gas saturation and recovery of gas by primary and secondary processes. In situ gas content measurements have been performed for 2 coalfields, i.e., Jharia coalfield and Raniganj coalfield from the wells under drilling. Most of the fields are still under exploration stage while a few have just started production.

References

Charles MB II, Bai Q (1998) Methodology of coalbed methane resource assessment. Int J Coal Geol 35(1–4):349–368

India's hydrocarbon outlook: annual report from the Ministry of Petroleum and Gas, India (2016–2017)

Levine JR (1993) Coalification: the evolution of coal as a source rock and reservoir rock for oil and gas. Am Assoc Pet Geol Stud Geol 38:39–77

Moore TA (2012) Coalbed methane: a review. Int J Coal Geol 101:36–81

Pashin JC (2011) Unconventional energy resources: 2011 review: natural resources research, coalbed methane. Am Assoc Pet Geol Energy Miner Div 20:282–286

Scott AR, Kaiser WR, Ayers WB Jr (1994) Thermogenic and secondary biogenic gases, San Juan Basin, Colorado and New Mexico-implications for coalbed gas producibility. AAPG Bull (Am Assoc Pet Geol); (US) 78(8), 1186–1209

Strąpoć D, Mastalerz M, Schimmelmann A, Drobniak A, Hedges S (2008) Variability of geochemical properties in a microbial dominated coalbed gas system from the eastern margin of the Illinois Basin, USA. Int J Coal Geol 76:98–100

Thakur P, Schatzel S, Aminian K (2014) Coal bed methane: from prospect to pipeline. Elsevier, Amsterdam

Tindal G (2013) Curriculum-based measurement: a brief history of nearly everything from the 1970s to the present. ISRN Educ 2013:29. Article ID 958530

Chapter 2
Geology of Probable Areas and Its Petrology

Abstract India has the fourth largest reserves of the coal in the world. Also, it's the 3rd largest producer of the coal in the world. This chapter targets the assessment of Gondwana and tertiary coal fields of India, their location in Indian sub-continent, its petro-graphical analysis etc. The following work will also elaborate India's activities in exploiting this unconventional hydrocarbon source.

Keywords CBM blocks India · Petrological analysis · CBM reserves · Coal evaluation

2.1 Geological Assessment of CBM Reservoirs

Indian coalfields are divided into two broad groups of two distinct geological ages: Gondwana coalfields of Permian age and Tertiary coalfields of Tertiary age. More than 98% of the Indian coal production comes from the Gondwana coals. These coalfields are prominent in the Damodar valley (West Bengal–Jharkhand), Son-Narmada valley (Madhya Pradesh), Mahanadi Valley (Orissa) and Pranhita–Godavari valley (Maharasthra–Andhra Pradesh). Raniganj and Birbhum coalfield in West Bengal; Bokaro, Jharia, North Karanpura and South Karanpura coalfield in Jharkhand; Singrauli, Sohagpur and Satpura coalfield in Madhya Pradesh; Ib-valley and Talchir coalfield in Orissa; Korba in Chattishgarh and Wardha in Maharasthra are the most important Gondwana coalfields having vast reserve of good quality coal. Tertiary coalfields are Assam–Meghalaya; Neyveli; Cambay; Barmer–Sanchor; Bikaner and Jammu & Kashmir. On the basis of coal rank, maturity, Physico-chemical attributes of coal, depth of occurrence of coal, available area and CBM potential, Indian coals are divided into four types: category-I, category-II, category-III and category-IV (Fig. 2.1). Jharia, Bokaro, North Karanpura and Raniganj coalfield belong to category-I type and comparable with global producing CBM fields in terms of gas content and adsorbed capacity (Hajra et al. 2003) for their good coal seam thickness, high rank and maturity. Damodar and Mahanadi valley coalfields are categorised as category-II and

S. Sharma et al., *Unconventional Resources in India: The Way Ahead*,
SpringerBriefs in Petroleum Geoscience & Engineering,
https://doi.org/10.1007/978-3-030-21414-2_2

category-III. The above mentioned Tertiary coalfields are placed in category IV as their CBM prospects are yet to be established.

India has started evaluating different coal-bearing sedimentary basins for their CBM potential in the early nineties (Patra et al. 1996). Exploration and development of coalbed methane in Jharia and Raniganj coalfield have been going on for about last 15 years and being actively explored by different exploration and production companies. India has a Prognosticated CBM Resource of around 92 (in TCF) of CBM reserve (DGH report 2016–2017) which is capable of fulfil filling the country's future growing energy demand up to a large extent. However, detail information of Indian CBM field is lacking in literature. Jharia coalfield of Jharkhand, Raniganj coalfield of West Bengal and Singareni coalfield of Andhra Pradesh have been selected for evaluation of chemical parameters, gas content, gas adsorption capacity, gas saturation and recovery of gas by primary and secondary processes. In situ gas content measurements have been performed for 2 coalfields, i.e., Jharia coalfield and Raniganj coalfield from the wells under drilling. Most of the fields are still under exploration stage while a few have just started production. According to the DGH report of 2017 the following production for the CBM for the financial year 2016–17 has been reported by major players (Table 2.1).

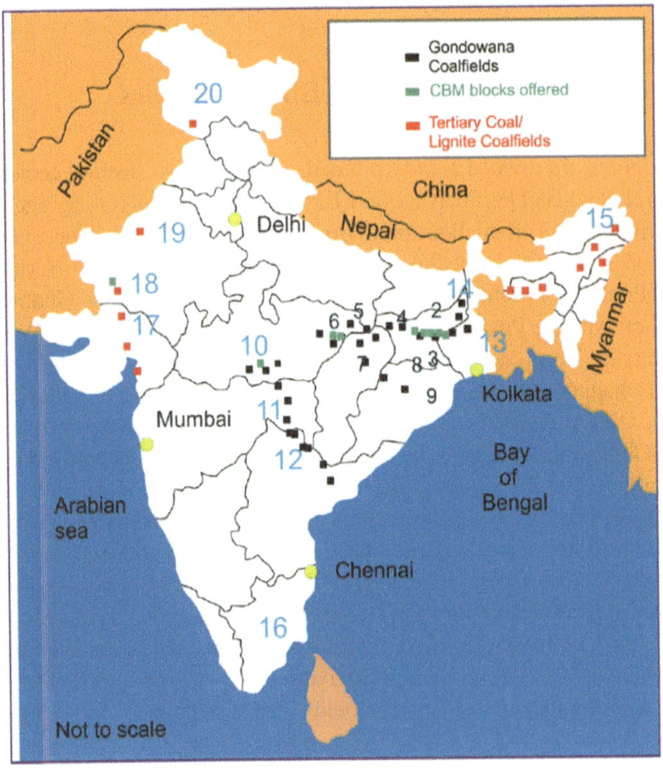

Fig. 2.1 Distribution of Coalfields and CBM blocks, India. *Source* DGH report (2016–2017)

Table 2.1 Major players for CBM resources in India

Industry	Region	Gas production (MMSCM)
ESSAR	Raniganj East	385.471
GEECL	Raniganj South	169.596
ONGC	Jharia	3.083
RIL	Sohagpur East/West	6.439
Total CBM production		564.589

Source DGH report (2016–2017)

Table 2.2 State-wise distribution of CBM resources in India

No.	State	Prognosticated CBM resource (in BCM)	Prognosticated CBM resource (in TCF)	Established CBM reserves (in TCF)
1	Jharkhand	722.08	25.5	1.916
2	Rajasthan	359.62	12.7	0
3	Gujarat	351.13	12.4	0
4	Odisha	243.52	8.6	0
5	Chhattisgarh	240.69	8.5	0
6	Madhya Pradesh	218.04	7.7	3.65
7	West Bengal	218.04	7.7	4.33
8	Tamil Nadu	104.77	3.7	0
9, 10	Telangana & Andhra Pradesh	99.11	3.5	0
11	Maharashtra	33.98	1.2	0
12	North East	8.50	0.3	0

Source DGH report (2016–2017)

According to the Directorate general of hydrocarbon (DGH report 2016–2017) as on March 2017, CBM production was estimated to be about 1.45 MMSCMD from the 4 CBM blocks. The estimates included the production from a CBM block in Jharia, Jharkhand being exploited by ONGC. Also The three major blocks Raniganj South, Raniganj East & Sohagpur West are being exploited for the production by the companies. At this current production rate it is estimated that the production rate for the financial year 2018–19 will be around 3.4 MMSCMD (Table 2.2).

2.1.1 *Different Petrological Analysis*

Laboratory Studies for CBM Prospect Evaluation

For the evaluation of a CBM prospect numerous laboratory studies needs to performed. This section will be dealing with the petrographic studies that are performed for the estimation of the resource in the exploration phase.

2.1.2 *Proximate Analysis*

Proximate analysis is done for studying the physiochemical parameters of the coal. The study is essential in establishing the ranks and grade for the coal. The chemical properties of the coal also play a significant role in estimating the gas sorption capacity of the coal. Eventually both the parameters are summed for estimating the gas content of the methane from the coal samples.

Estimation of moisture: A known amount of sample (Powdered and dried) should be uniformly distributed onto the petridish. The petridish should be preheated above 100 °C and cooled and weighed. Now the combined weight of the uncovered petridish and the sample is noted. The system is heated above 100 °C ± 20 °C form about 2 h. The heating should be stopped as soon as there is no further loss in the mass of the system. The petridish should be covered and allowed to be cooled in the desiccator. The system should be weighed again. A loss in mass of the system is observed after the drying process. The loss in mass of the system should be expressed as the total moisture content of the sample.

Estimation of ash: A known amount of sample (Powdered and dried) should be taken in a clean and dried crucible. The crucible should be weighed prior to the experimentation. The crucible and sample should be weighed together. The sample needs to be inserted into the muffle furnace. The system should be heated at about 500 °C in the presence of air for about half an hour. The temperature of the muffle furnace should be raised to about 800 °C ± 10 °C in the muffle furnace and should be maintained until no further reduction in mass is observed. The crucible containing the sample should be taken out from the muffle furnace. The sample should be covered and cooled in the desiccator. The cooled sample should be weighed. The mass of the ash obtained from the sample is reported as the mass% of the total mass.

Determination of volatile matter (VM): A known amount of sample (Powdered and dried) should be taken in clean crucible. The weight of the crucible is taken prior filling it with sample. The crucible containing the sample is weighed and kept in the muffle furnace. The sample is heated in the absence of air at about 900 °C ± 10 °C for approximately 7–8 min. The crucible covered with the lid is taken out from the furnace. The sample is then kept in the desiccator for cooling. The cooled sample is weighed. The percentage reduction in mass from the original mass is reported as the VM contained in the sample.

Determination of fixed Carbon: Percentage of fixed carbon can be estimated by taking the difference fraction of moisture, ash and volatile Matter from unity. It can be reported as the percentage also.

Calculation of vitrinite reflectance (R_o %): The magnitude of the (R_o %) helps in the characterization of the rank of coal. The R_o % can be estimated with the help of an equation suggested by Rice. This equation utilises the data obtained from the proximate analysis. The equation is as follows:

$$R_o\,\% = -2.712 \times \log(\mathrm{VM}) + 5.092$$

where, 'R_o' is reflectance value of vitrinite macerals and 'VM' is the dry, ash free volatile matter of coal (Rice 1993).

References

Hazra PN, Rudra M, Guha S, Kar MK, Basumatary JK, Kumar A (2003) Geochemical characterization of coalbed gas of Jharia & Raniganj Basins and its implications. In: Proceedings of the international conference. Mussoorie, India

India's hydrocarbon outlook: annual report from the Ministry of Petroleum and Gas, India (2016–2017)

Patra TC, Pandey AK, Dutta HC (1996) Potential areas of coal seam methane in India. J Geol Soc India 47:215–227

Rice DD (1993) Hydrocarbons from coal. AAPG studies in geology, vol 38. American Association of Petroleum Geologists, Tulsa, OK, pp 159–184

Chapter 3
Gas Content Measurement in Coalbed: Desorption Test and Isotherm Study

Abstract Coal has always been associated with the generation of hydrocarbon gases (mostly methane). However, the production of gases on a commercial scale has been the aim of the industry. The assessment of gas content of coal is an important parameter which helps in the estimation of total gas content of the coal. This chapter illustrates the different methods available for the estimation of gas content of coal. Also, it elaborates the gas content of the Indian coal fields.

Keywords Gas content · Isotherm studies · Methods of evaluation

3.1 Introduction to Different Methods Used

Estimating the volume of methane contained in the coal seams is an important parameter for potential gas content. An early estimation helps in calculation of gas resource and as an input parameter for simulation jobs. Simulated results helps in forecasting the gas production and control option for the generated gas.

Coal is a microporous structure of organic origin. The microporous structures are stacked one over the other. They have natural fractures of cleat system that runs mostly perpendicular to each other. These structures are attributed to the coalification process along with the tectonic stress that may have been applied. The face cleat and fractures are the dominant structures responsible for majority of the permeability of the coal seams. Gas adsorbed in the coal seams is different from the ones found in conventional reservoirs. It is stored as the free gas in the cleat structure or it is adsorbed on the inner surfaces of the micro-pores present in of the coal. Scientist have described that the gases present in the coal bed are adsorbed on the surface of the coal in a near liquid form. Whereas in conventional reservoirs the gas is stored as free gas in the pores. The different storage mechanism of the gas in coal seams make the conventional techniques incapable of estimating the gas content within it. Thus alternate techniques have been proposed for the estimation of volume of the gas adsorbed inside the matrix of coal seam (Diamond and Schatzel 1998; Rice 1993).

S. Sharma et al., *Unconventional Resources in India: The Way Ahead*,
SpringerBriefs in Petroleum Geoscience & Engineering,
https://doi.org/10.1007/978-3-030-21414-2_3

Coal bed methane gas is estimated primarily for two reasons: Miners safety and the amount of total gas in place. The estimation of the gas content from the coal is done by two techniques (1) Direct measurement (2) Indirect Measurement. The direct measurement technique focuses on the estimation of the volume of the gas released from the coal under consideration in an isolated desorption container. Whereas, indirect method estimates of the volume of the desorbed gas by utilizing different empirical isotherms developed in the laboratory.

The most common technique for the estimation of gas content from the coal sample has been divided into three sections: determination of lost gas, adsorbed gas and the residual gas. These sections are individually estimated by using different procedures. The estimation of the desorbed gas from the different sections are utilized to calculate the total gas retrieved from the coal sample. Before the estimation technique we should introduce the terms that will be used for the estimation of the total gas content (Sawyer et al. 1987; Owen and Sharer 1992).

Lost gas: It is reported to be the portion of gas that escapes from the sample prior to its containment into the desorption canister and sealed for further test. It has been a difficult task to estimate the volume of lost gas directly. Indirect method from the gas desorption test have been utilized for its estimation. The amount of the lost gas is found to be dependent upon numerous factors. The primary factors can be named as the physical properties of the sample, type of drilling fluid used for the operation, retrieval time of the sample and water saturation in the sample. Most of the methods are dependent upon extrapolating the desorption data plotted over the time period.

Determination of lost gas time plays a vital role when the commercial aspects are associated with it. Lost gas time are dependent upon the depth of the well from which the cores are being retrieved using wireline operations. For commercial Indian fields where the depths of the well are ranging from 1000 to 1500 m. In these formations the average lost gas time is greater than one hour. These increased lost gas time can significantly contribute to the errors in the estimation of volume of lost gas.

Many researchers have reported that, the estimation of the lost gas is also dependent upon the physical character of the retrieved sample. In reality the physical character governs the rate of desorption rate of the gas from the coal sample, thereby affecting the content of lost gas in the system. The consolidated sample are prone to less desorption of gas as compared to the brittle ones. The brittle sample are found to exhibit a larger desorption rate attributing to the increased surface area. It has also been observed that the density of the drilling fluid being used also affects the desorption rate. This happens because the drilling fluid affects the estimation of time of initialization of the desorption. The hydrostatic pressure controls the process.

Desorbed gas: Its estimation starts with the isolation and retrieval of coal sample into the desorption canister. The gas being desorbed from the sample starts to accumulate into the canister. This facilitates the direct measurement of the desorbed gas from the coal. The volumetric method using the fluid displacement are most common method utilised for its estimation. A probable setup has been developed and reported by the researcher Ojha et al. Desorption canister used in the apparatus is shown in "a" of figure. Whereas the part b utilizes a volumetric balance system which measures the volume of desorbed gas by measuring the change in height of the fluid column with time (Fig. 3.1).

Considerably more accurate methods have been developed by the different government agencies to evaluate the amount of gas desorbed from the samples. These methods are based on the calculation of differential pressure in the canister as the gas is released. Further, they have been found to utilize the ideal gas law for estimating the amount of desorbed gas (Ojha et al. 2013; Thakur et al. 2014).

Residual gas: It has been experimentally observed that the desorption rates of the gas from the coal sample declines continuously. Researchers have reported that, as the excessive low desorption rates are achieved the experimental observations can be stopped. Desorption rate can vary from days to months depending upon the type of coal (friable, blocky etc.). Even at these rates there may be certain amount of CBM left trapped inside the structure of coal. The trapped gas is reported to be retrieved by crushing the sample from the previous step in an enclosed container. The other procedure remains the same as canister desorption test.

Total gas content

It is reported as the cumulative of lost gas (V_{Lost}), desorbed gas ($V_{Desorbed}$) and residual gas ($V_{Residual}$). It is reported in cubic centimetre/gram or metre^3/ton. The total volume of gas obtained per unit weight of coal was estimated by the addition of V_{Lost}, $V_{Desorbed}$, and $V_{Residual}$ divided by the total weight (W) of the sample.

$$\text{Gas content,} \quad V_T = \frac{V_{Lost} + V_{Desorbed} + V_{Residual}}{W}$$

Fig. 3.1 Canister and Desorption apparatus set-up

3.2 Indirect Method of Gas Capacity Estimation

It has been reported that the majority of the gas in coal is found to adsorbed on the internal surface of microspores. The amount of gas adsorbed varies directly with pressure and inversely with temperature. The correlation relating the volume of adsorbed gas with pressure and temperature has been found to be dependent upon the moisture and ash content of coal samples. The correlation also known as Kim's empirical equation utilized for the estimation of gas capacity is as follows (Kim and Douglas 1973).

$$G_{saf} = (0.75)(1 - a - w_c)\left[K_o(0.095d)^{n_o} - 0.14\left(\frac{1.8d}{100} + 11\right)\right]$$

$$K_o = 0.8\frac{x_{fc}}{x_{vm}} + 5.6, \quad n_o = 0.315 - 0.01\frac{x_{fc}}{x_{vm}}$$

where,

G_{saf} Dry ash free gas storage capacity, cm^3/g;
a Ash content, weight fraction
w_c moisture content, weight fraction;
d Sample depth, m
x_{fc} Fixed carbon, weight fraction;
x_{vm} Volatile matter, weight fraction.

Also, numerous empirical correlations of the measured gas content are plotted against numerous variables. The major variables are coalbed depth and coal rank. These plots can be used for the estimation of gas content of the coal. These laboratory estimation of the gas content using different isotherms still fails to provide the in situ gas content accurately. The inaccuracy is attributed to the quantitative measurement of the parameter at varying temperature and pressure conditions. One of the most common isotherm method is the Langmuir Isotherm. It helps in estimating the storage capacity of the gas. The isotherm generation is performed on crushed samples. The sample under consideration must be fresh as oxidation reduce the storage capacity of the sample. It also makes it difficult to estimate correct moisture content of the sample. The gas storage capacity can be estimated using the Langmuir isotherm. It is given by the expression

$$G = (1 - f_a - f_m)\frac{V_L P}{P_L + P}$$

where,

G Storage capacity of gas, scf/ton
f_a fraction of ash in coal sample
f_m fraction of moisture in coal sample
P Reservoir Pressure, psia

V_L Langmuir volume constant, scf/ton
P_L Langmuir pressure constant, psia.

Here the Langmuir volume is defined as the maximum volume of gas that a coal can absorb on its surface (internal/external). Whereas, the Langmuir pressure is represented as the pressure at which the half of the Langmuir volume can be desorbed. In case of CBM the Langmuir pressure is also known as critical desorption pressure. The diagram below illustrates the estimation of different parameters required for the estimation of gas content of the coal (Fig. 3.2).

These isotherms methods of estimation tend to overestimate the gas present in the structure of the coal. This is attributed to the assumption the coal seams are completely saturated with the gas. However, this is not true everywhere, especially in case of shallow reservoirs where the coal seams are not completely saturated with the gas (Busch and Gensterblum 2011; Rice 1993).

3.2.1 *Sorption Time*

Sorption Time denoted by (τ) is an estimate of the rate of diffusion of the CBM from the coal sample. It is reported as the time required to desorb 63.2% of the initial volume of the gas present in the sample. Careful estimation helps in proper estimation of the reserves and gas production rates.

Sorption time can also be calculated using cleat spacing and diffusion coefficient by the expression $\tau = \frac{S^2}{8\pi D}$.

Where, "τ" is the value of sorption time, "S" is the cleat spacing in ft and diffusion coefficient "D" in ft^2/day. Estimation of the above mentioned parameter in the expression is a tedious task. Hence the overall property, sorption time (τ) is determined using the CBM content measurement which is directly used as input parameter for reservoir simulation (Sawyer et al. 1987).

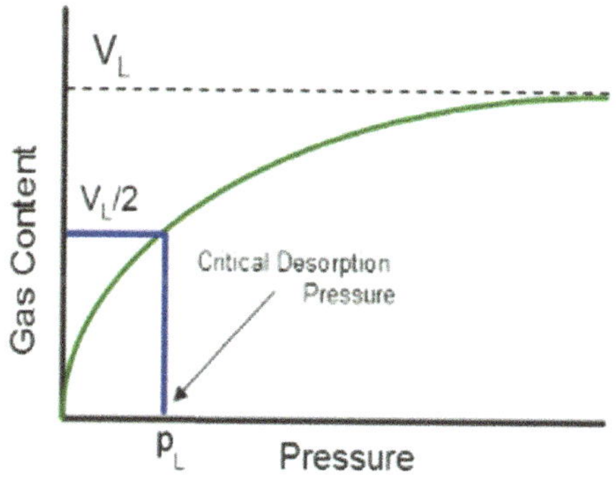

Fig. 3.2 The Langmuir isotherm curve used for the CBM reservoirs

3.2.2 Indian Status of CBM Gas Content

Researchers from India have reported that is-situ gas content of coal samples from the Jharia and Raniganj fields have the gas contents in the range of 4.81–25.43 cm^3/g as dry ash free basis (daf) and 4.42–4.52 cm^3/g (dry ash free basis) respectively. The range of gas content for Jharia coalfield fall within and above the range of economic limit and are targeted as suitable zone for CBM extraction buy the players like ONGC and Essar. Also the maximum adsorption capacity of samples from Jharia coalfield varies from 12.20 to 23.00 cc/g (daf), and the degree of gas saturation varies from 34.90% to near saturation (95.13%). The critical desorption pressure (CDP) is reported to be less than the reservoir pressure for most of the cases. So, dewatering period is required before initial gas production from these areas. Predicted production characteristics are seen to behave similarly as observed from those of nearby development wells (Ojha et al. 2013).

References

Busch A, Gensterblum Y (2011) CBM and CO_2-ECBM related sorption processes in coal: a review. Int J Coal Geol 87:49–71

Diamond WP, Schatzel SJ (1998) Measuring the gas content of coal: a review. Int J Coal Geol 35:311–331

Kim AG, Douglas LJ (1973) Gases desorbed from five coals of low gas content, U.S. Bureau of Mines, Washington, DC, Report of Investigations 8043, p 22

Ojha K, Mandal A, Karmakar B, Pathak AK, Singh AK (2013) Studies on estimation & prospective recovery of coal bed methane from Raniganj Coalfield, India. Energy Sources Part A Recover Util Environ Eff 35:426–437

Owen LB, Sharer J (1992) Method calculates gas content per foot of coalbed methane pressure core. Oil Gas J 2:47–49

Rice DD (1993) Hydrocarbons from coal. AAPG studies in geology, vol 38. American Association of Petroleum Geologists, Tulsa, OK, pp 159–184

Sawyer WK, Zuber MD, Kuuskraa VA, Horner DM (1987) Using reservoir simulation and field data to define mechanisms controlling coalbed methane production. In: Proceedings of the 1987 coalbed methane symposium. The University of Alabama, Tuscaloosa, pp 295–307

Thakur P, Kashy S, Aminian S (2014) Coal bed methane, 1st edn. Elsevier

Chapter 4
Recovery of Methane and CO_2 Sequestration

Abstract Recovery of CBM on a commercial scale is associated with the dewatering process. The release in pressure liberates the adsorbed gas for the production. The vacant site can be utilized for the adsorption of other gases. Also the incumbent gas molecules can be displaced with a gas molecule that has more affinity towards the coal as compared to that of methane. This chapter addresses the mechanism and the produced water disposal problem associated with the CBM production. It also elaborates the measures taken in India for the management of produced water.

Keywords CBM production · CO_2 sequestration · Dewatering · Water disposal

4.1 Introduction

Currently India is focusing on the recovery and utilization of CBM on a commercial scale. It has been observed that as the initial recovery rates of the CBM production decline after few years of the start of production. This affects the overall economics of the operation. Hence, Carbon dioxide (CO_2) is being injected into the formation for enhancement of the production of CBM. This serves a dual purpose of reducing the greenhouse gas in atmosphere along with the utilization of unconventional hydrocarbon resource. Although there are numerous technological complexities associated with this recovery technique like exploitation of heterogeneous formations, hydro fracking operations etc.

It has been reported by numerous scientist that coal can adsorb large volume of CO_2 as compared to that of CBM. However, the formation needs to be assessed for consequences prior to CO_2 injection. Several lab and field studies have indicated that sequestration of CO_2 to produce CBM results in the swelling and shrinkage of the coal seam. Burlington Resources have been utilizing the CO_2 Injection practise and have subsequently enhanced the CBM recovery from the conventional reservoirs. This also helps in long term disposal of CO_2.

S. Sharma et al., *Unconventional Resources in India: The Way Ahead*,
SpringerBriefs in Petroleum Geoscience & Engineering,
https://doi.org/10.1007/978-3-030-21414-2_4

The Indian CBM reservoir of the Gondwana region are found to be heterogeneous in nature. The heterogeneity plays a vital role in the selection of areas for the application of CO_2 as an enhanced methane recovery option (Mendhe et al. 2017a, b, 2018).

4.1.1 *Adsorption of CO_2 and Methane*

It has been reported that the methane adsorption capacity and be utilized for the estimation of adsorption capacity of CO_2. Vitrinite reflectance (%) and the proximate analysis of the coal samples are the major parameters affecting the adsorption capacities of the gases. When CO_2 is injected into the coal it can be adsorbed into the microspores present in the structure or occupy the surface area of the coal. CO_2 has been found to get adsorbed onto the coal even by displacing the methane adsorbed on the surface. This phenomenon is attributed to the fact that CO_2 has more affinity towards coal. It is found that two/three molecules of CO_2 can displace and occupy the adsorption site of a methane molecule. However, studies have indicated that this adsorption ratio is dependent upon the rank of coal. This ration is approximately two for bituminous coal and it reaches around ten for low ranks coal. It has been found that the pressure also affects the adsorption ratios. It has been reported that at higher pressures around 1450 psi the CO_2 passes into the supercritical state thereby increasing the adsorption ratios. Hence, it is important to understand the adsorption mechanism of both CO_2 and CBM for its successful implementation in the field (Mendhe 2017a, b; Vishal 2017).

4.2 Potential Sites for Adsorption

Indian Coalbeds are classified as grey and concealed. It is because of the depth at which they are found and its grade characteristics. It has been reported that Indian Subcontinent hold the storage capacity of about 5 Giga tonnes of CO_2 in its unmineable coal seams. As a convention the coal seam below the depth of 6500 ft are not considered for CBM production of CO_2 Sequestration. Raniganj, Barakar and Jharia regions of Indian subcontinent contains a good quality of coal. Presently these are the major regions for commercial CBM production from the coal seams. These areas hold a major potential for using CO_2 as a tool to enhance CBM production rates. This is because of the fact that the following areas have an easy access for the CO_2 entrapment from the nearby power plants. According to the studies performed by the Holloway, and IEAGHG there is a huge potential of about 6000 billion cubic feet (bcf) of CO_2 storage. This is because the major Indian fields for CBM production are found to be in the range of 4000 ft. Where Mendhe and fellow researchers have reported a CO_2 storage of the about 80,260 bcf in the Gondwana and tertiary coals of India. Indian power sector is largely dependent upon the coal as a source of power generation. Hence there are ample amount of power and other associated industry that can serve as a point source to aid CO_2 sequestration and flooding in the coal seams. The Cambay basin is marked with a sequestration potential of about 1885 Metric Tonne. Whereas, some regions of Singrauli have shown a potential of about 2 Metric tonne (Mendhe et al. 2017a, b; Ojha et al. 2013; Singh et al. 2018).

4.3 Water Production Problem

CBM wells are associated with a mass production of water produced during the dewatering phase of the well. A typical CBM well releases the adsorbed methane on the surface of the coal once its depressurized by the removal of water. A typical CBM well follows the sequence given in the figure for its course of production (Fig. 4.1).

The water being produced from the CBM well is one of the major concern associated with this clean energy source. The water produced alongside of CBM can't be reinjected into the formation. However, it is disposed or treated to reduce the total dissolved solid (TDS) before it can be implemented for different usages. The disposal remains a complicated task owing to the poor quality of water produced from the wells. The high cost of disposal per barrel of water produced is one of the major restriction for the small operators to invest in this clean energy source. The factors attributing to the elevated cost are its pipeline maintenance, electrical & maintenance cost for pumps, working life of the facility, injection well depths, water treatment cost before disposal etc. (Singh et al. 2018; Mendhe et al. 2017a, b).

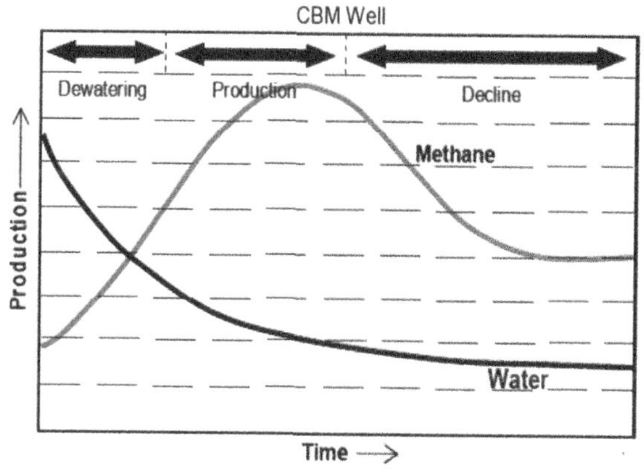

Fig. 4.1 Timeline for the CBM well production

4.4 Produced Water Management Practised in India

With the quality of water being produced from the Indian oil fields, following methods might be appropriate to treat and dispose it. Irrigation is among a probable solution for the disposal of CBM water. Prior to implementation the water should follow desalinization and suitable treatment. The produced water can be impounded into storage facilities, reservoirs ponds etc. The solution is being used for the management of water that is being produced from the Raniganj coal fields, India. It has been found to be effective up till now for that particular region. However, the drinking water still remains a major concern here. Further for making the water good for household usage it further needs to be treated with reverse osmosis (RO). The major problem emphasized for these field are of desalination, de-gassing, harmful compound removal (organic compounds, heavy metals, carcinogenic compounds, biological compounds) etc. The technologies used for the treatment in the above fields include removal of TDS by precipitation, RO, oxidation (electro-chemical or photolytic), aeration, gravity settling, coagulation, flocculation etc. (Singh et al. 2018; DGH report 2016-2017).

4.5 Conclusion

The following conclusion can be drawn from the details being discussed in the CBM sections

- India has a Prognosticated CBM Resource of around 92 (in TCF) of CBM reserve.
- Jharia coalfield of Jharkhand, Raniganj coalfield of West Bengal and Singareni coalfield of Andhra Pradesh are among the major CBM fields of India.
- Indian coals are divided into four types: Category-I, category-II, category-III and category-IV.
- CBM production was estimated to be about 1.45 MMSCMD from the 4 CBM blocks.
- The gas content from the Indian coal can be estimated using Langmuir or Kim's correlations with a substantial accuracy.
- Dewatering period required before initial gas production from Indian Coal fields is considerably low.
- The Indian coal has a very large CO_2 storage capacity which can help in enhancing the recovery of CBM and reducing the greenhouse gas.

References

India's hydrocarbon outlook: annual report from the Ministry of Petroleum and Gas, India (2016–2017)

Mendhe VA, Kamble AD, Bannerjee M, Mishra S, Sutay T (2017a) Coalbed methane: present status and scope of enhanced recovery through CO_2 sequestration in India. Springer, Green Energy and Technology (Chapter 13)

Mendhe VA, Mishra S, Varma AK, Kumar A, Singh AP (2017b) Coalbed methane-produced water quality and its management options in Raniganj Basin, West Bengal, India. Appl Water Sci 7:1359–1367

Mendhe VA, Mishra S, Varma AK, Kumar A, Singh AP (2018) Management of coalbed methane and coal mine produced water for beneficial use in Damodar Basin of India. Water Resour Manag 78:283–296

Ojha K, Mandal A, Karmakar B, Pathak AK, Singh AK (2013) Studies on estimation & prospective recovery of coal bed methane from Raniganj Coalfield, India. Energy Sources Part A Recover Util Environ Eff 35:426–437

Singh U, Singh AK, Singh DB (2018) Coalbed methane-produced water characteristics and insights from the Jharia coalfield in India. Energy Sources Part A Recover Util Environ Eff 40 (16):1897–1909

Vishal V (2017) Recent advances in coal seam sequestration research in India—highlighting multiphase CO_2 flow for deep seam sequestration. Energy Procedia 114:5377–5380

Part I
Coal Bed Methane (CBM): Conclusion

The following conclusion can be drawn from the details being discussed in the CBM sections

- India has a Prognosticated CBM Resource of around 92 (in TCF) of CBM reserve.
- Jharia coalfield of Jharkhand, Raniganj coalfield of West Bengal and Singareni coalfield of Andhra Pradesh are among the major CBM fields of India.
- Indian coals are divided into four types: Category-I, category-II, category-III and category-IV.
- CBM production was estimated to be about 1.45 MMSCMD from the 4 CBM blocks.
- The gas content from the Indian coal can be estimated using Langmuir or Kim's correlations with a substantial accuracy.
- Dewatering period required before initial gas production from Indian Coal fields is considerably low.
- The Indian coal has a very large CO_2 storage capacity which can help in enhancing the recovery of CBM and reducing the greenhouse gas.

Part II
Shale

Preface

Unconventional gas or oil is very specific type of hydrocarbon that needs exploration methods which are unique to the traditional ones. Shale oil and gas is one type. The first chapter in the book relates to introducing what are shales, their characteristics, origins, types and how they contribute to energy.

The second chapter provides an idea for global scenario in shale reserves. Shale has the highest contribution in boom in energy sector in USA. Similar can happen in many other countries such as China and Argentina with technology development.

The third chapter is focussed entirely on India, its shale reserves, and growing potential with shale growth.

Unconventional resources are the need of the time, but they pose greater environmental challenge. The last chapter in the book has been dedicated to the problems or difficulties in shale exploration.

Chapter 5
Introduction to Shale

Abstract Today's energy intensive lifestyle needs exploration of hydrocarbon such as crude oil and natural gas. Hence, there is a need to develop some unconventional resources such as shales, gas hydrates and CBM. They are called unconventional due to their need for new and different technology for exploration. This chapter is dedicated to introduction of one of the unconventional resource i.e. shales, their composition, structure and uses. The first exploration of shale gas was as early as in 1825. As the present exploration of shale will have a major impact on future energy needs, it becomes very necessary to know about the shale resources and its characteristics.

Keywords Shale origin · Shale composition · Permeability

5.1 What Are Shales?

Shales are fine grained sedimentary rocks that are formed from the compaction of silt and clay. The clay present is formed from the decomposition of mineral feldspar; Quartz, Mica, pyrite and organic matter are also present. More than 60% of world's sediments comprise of shales. The thickness of shale layer vary from 0.5 mm to 30 cm. As the sand, silt and carbonate content of shale increases, its thickness also increases. Conversely, when clay and organic content increases, the thickness of shale decreases. The stratification (layering) in shales depends on many factors like rate of sedimentation, composition, compaction state and grain size (Potter et al. 2012).

5.2 Type of Shales

The shale rock act as both source and reservoir rock i.e. both as creator and storer of H/C. Shales gas are formed by two methods; Biogenic shale gas is formed by bacterial decay of microbes and thermogenic shales are formed by thermal cracking

S. Sharma et al., *Unconventional Resources in India: The Way Ahead*,
SpringerBriefs in Petroleum Geoscience & Engineering,
https://doi.org/10.1007/978-3-030-21414-2_5

of oil into gas. Based on their content and origin, shales are of two types; Marine shales are brittle in nature and low in clay content. Clay content is found to be higher in non-marine shales (Zendehboudi and Bahadori 2017). Based on content, shale is also categorized in below categories:

- Carbonate rich shales which contains substantial amount of carbonates such as calcite or dolomite.
- Siliceous shale: These are usually dark brown or black shales and have minerals such as Quartz, feldspar or clay as its main constituents.
- Cannel Shale: It majorly contains organic matter. Colour is dark brown or black. It has high amount of algae remains (Chilingarian and Yen 1976).

Origin of Shale: Shale results from continuous deposition of mineral debris and organic degradation. Prior development of anaerobic conditions and abundant organic productivity are needed for formation of shales. Overburden pressure coupled with continued sedimentation is necessary for digenesis. If there is chemical activity at low temperature, it results in loss of volatile fractions leading to formation of sedimentary rock with high content of organic residues.

5.3 Characteristics of Shales

Conventional gas reservoirs are created when natural gas migrates toward the Earth's surface from an organic-rich source formation into highly permeable reservoir rock, where it is trapped by an overlying layer of impermeable rock. In contrast, shale gas resources form within the organic rich shale source rock. The low permeability of the shale greatly inhibits the gas from migrating to more permeable reservoir rocks. Without horizontal drilling and hydraulic fracturing, shale gas production would not be economically feasible because the natural gas would not flow from the formation at high enough rates to justify the cost of drilling.

Shale is different from conventional reservoirs as conventional resources are formed by migration followed by their trapping in impermeable cap rock while shale resources are formed within the Shale source rock (US Department of Energy 2010). Shale gas has lowest permeability of about 0.0001 mD (Rahim and Al-Anazi 2012). Porosity is most of the time below 10% (www.Geology.com; Speight 2012b). The thickness range for productive portion of shale is 2–5 m (Speight 2012b; Martini et al. 1998; Speight 2012a). Typical depth for shales varies from 1000 to 5000 m. Shales are characterized by the low ratio of mineral to organic material content. Total organic content in shales are usually above 1%.

5.4 Composition of Shales

Shale gets deposited in shallow lake, marshes or seas, small lakes, bogs and lagoons. The large lake basins contain the thickest and some of the richest oil shales. Shales has over 33% of ash content. Usually shale comprise of clay, silica, carbonate and organic material where kerogen, quartz, clay, carbonate and pyrate are present as primary component and Uranium, Iron, Vanadium, Nickel and Molybdenum as secondary component. According to colour, there are black (dark) shale and light shales. Light shales have less organic content while black shales have large organic content (Zendehboudi and Bahadori 2017). Gas is stored in shales via 3 ways: as free gas i.e. gas within the rock pores and natural fractures; as adsorbed gas i.e. when gas is adsorbed on organic material or when gas is dissolved (Mailluri 2011).

5.5 Applications of Shales

Shales can be used in electricity production, industrial use, cooking and heating at home etc.

References

Chilingarian GV, Yen TF (1976) Developments in petroleum science, vol 5, Oil shale, Elsevier Scientific Publishing Company. ISBN 0-444-41408

Geology.com, http://geology.com/rocks/shale.shtm

Mailluri G (2011) Gas shale reservoir: characterization and modelling play shale scenario on wells data base. ENI Corporate University, San Donato Milanese, Italy

Martini AM, Walter LM, Budai JM, Ku TCW, Kaiser CJ, Schoell M (1998) Genetic and temporal relations between formation waters and biogenic methane: upper Devonian Antrim Shale, Michigan Basin, USA. Geochim Cosmochim Acta 62(10):1699–1720

Potter PE, Maynard JB, Pryor WA (2012) Sedimentology of shale, study guide and reference source. Springer, New York. ISBN-13: 978-1-4612-9983-7

Rahim Z, Al-Anazi H (2012) Improved gas recovery—1: maximizing postfrac gas flow rates from conventional, tight reservoirs. Oil Gas J

Speight JG (2012a) Shale oil production processes, 1st edn. Gulf Professional Publishing. eBook ISBN 9780124017498

Speight JG (2012b) Shale oil production processes, 1st edn. Gulf Professional Publishing. eBook ISBN 9780124045514

US Department of Energy (2010) http://www.eia.doe.gov/energy_in_brief/about_shale_gas.cfm, Dec 2010

Zendehboudi S, Bahadori A (2017) Shale oil and gas handbook. Gulf Professional Publishing. ISBN 9780128021002

Chapter 6
Global Scenario

Abstract Of all the substitute of petroleum reported this far, shale stands out to be of high importance. Shale has been an important source of energy for Australia, France and Scotland. As far as US is concerned, over 700 billion barrels of synthetic crude spread of over 11 million acres of Colarado, Uthah and Wyoming District. Only about 2% of whole 30 trillion bbl of shale oil is available for commercial exploration. Total recoverable natural gas in the world is 28,205 tcf of which shale gas is almost one-fourth.

Keywords Shale reserves · Production from shale

6.1 World Reserves & Production of Shale

Globally available resources of shale gas are approximately 456×10^{12} m^3. By 2035, shale gas will have 50% share in total energy produced by USA which is approximately 340 billion cubic meters/year. Most of the shale reserves of the world are found in USA, Canada, China, Australia, and India (Zendehboudi 2017). In decreasing order, the shale reserves are prominent in China, Argentina, Algeria, USA and Canada (Zendehboudi 2017). At CAGR of 4.7%, global shale production rate is expected to grow from 5563 billion cubic feet in 2016 to 69,991 billion cubic feet in 2021. Good growth rate from 2016 to 2021 is expected from China (6.2%), Poland (6%), France (5.4%), South Africa (5.1%), and the U.S. (5%). With 78% share, North America will remain the largest shale gas producer worldwide by the year 2021. It will be followed by EMEA (13%), Asia Pacific (7%), and ROW (2%). The growth in these regions is due to technological expertise and availability of resources (Tables 6.1 and 6.2).

S. Sharma et al., *Unconventional Resources in India: The Way Ahead*,
SpringerBriefs in Petroleum Geoscience & Engineering,
https://doi.org/10.1007/978-3-030-21414-2_6

Table 6.1 Shale gas reserves (U.S. Energy Information Administration 2015)

Rank	Country	Amount of shale gas (trillion cubic feet)
1	China	1115
2	Argentina	802
3	Algeria	707
4	USA	665 (1161)
5	Canada	573
6	Mexico	545
7	Australia	437
8	South Africa	390
9	Russia	285
10	Brazil	245
	Total	7299 (7795)

Table 6.2 Proved and unproved gas resources in the world (U.S. Energy Information Administration 2015)

USA	Amount of wet natural gas (trillion cubic feet)
Shale gas proved reserves	97
Shale gas unproved reserves	567
Other gas proved resources	220
Other gas unproved reserves	1546
Total	2431
Increase in total gas resources due to shale gas	38%
Share of shale gas in total	27%
Outside USA	
Shale gas unproved reserves	6634
Other gas proved resources	6521
Other gas unproved reserves	7269
Total	20,451
Increase in total gas resources due to shale gas	48%
Share of shale gas in total	32%
Total world	
Shale gas proved reserves	97
Shale gas unproved reserves	7201
Other gas proved resources	6741
Other gas unproved reserves	8842
Total	22,882
Increase in total gas resources due to shale gas	47%
Share of shale gas in total	32%

6.2 United States of America (USA)

USA is at top in world shale gas production and accounts for 40% of its natural gas production. USA saw a major breakthrough in production of shale from almost negligible in 2000 to 10 bcfd in 2010. Major reserves of shale in US are Barnett, Haynesville, Fayetteville, and Marcellus. As per the projection from Annual Energy Outlook, production of shale gas is expected to constitute 45% of U.S. total natural gas supply by 2035 (US Department of Energy 2010).

6.3 Canada

Canadian government is now focussing more on shale gas production. Still, the commercial production from shale reserves is yet to start. Canadian gas industry is also changing with continued increased focus on shale gas production. Potential areas are British Columbia, Alberta, Ontario, Quebec and Mauritius.

References

Analysis & Projections, http://www.eia.gov/analysis/studies/worldshalegas/

Annual Energy Outlook (2015) https://www.eia.gov/todayinenergy/detail.php?id=25832

US Department of Energy (2010) http://www.eia.doe.gov/energy_in_brief/about_shale_gas.cfm, Dec 2010

Zendehboudi S, Bahadori A (2017) Shale oil and gas handbook, theory, technologies and challenges. ISBN 978-0-12-802100-2

Chapter 7
Shale Resources in India

Abstract Shale has been a promising source of hydrocarbon augmenting the increasing dependence of the country on these fossil fuels. The current Indian scenario has a dependence of around 80% on imports to cater its demand for the fossil fuel. India imports 80% of its oil requirement, and its domestic oil production has also fallen continuously for six years in a row. The fall is due to lack of efficient drilling from matured oil fields. The technological advancements in the form of hydraulic fracturing have made the exploitation from the shales an economic prospect. The chapter deals with the brief introduction to the prospective basins for the shale exploitation in India.

Keywords Shale India · Shale basin

7.1 Indian Scenario for Shale

India's energy demand as a percentage of global energy demand is expected to rise to 11% in 2040 from 5.58% in 2017. Country's crude oil consumption is expected to grow at a CAGR of 3.6% to 500 million tons by 2040 from 221.76 million tons in 2017. As per the annual report (2017) of the Directorate General of Hydrocarbons (DGH) the Indian shale gas reserves are estimated to be in the range of around 8.5 trillion cubic meter (TCM) to 59.5 TCM. However, for shale oil India has explored the basins Cambay, Krishna Godavari (KG), Cauvery and Indo-Gangetic basins, Damodar Valley basins. The DGH has reported some other potential reserves like Upper Assam, Vindhyan, Pranhita-Godavari and South Rewa. Indian oil major, Oil and Natural Gas Corporation (ONGC) is the prime player for the exploration of shale reserves. In October 2013 ONGC has drilled the first exploratory well for the shale gas exploitation in the Jambusar region of Cambay basin. From this well it was estimated that the basin holds around 20 TCF of recoverable shale gas. Up till now ONGC has drilled 22 wells for the collection of core samples. Extensive studies are being carried out on these samples for assessing the commercial viability of the resource with respect to Indian scenario (DGH report 2017).

S. Sharma et al., *Unconventional Resources in India: The Way Ahead*,
SpringerBriefs in Petroleum Geoscience & Engineering,
https://doi.org/10.1007/978-3-030-21414-2_7

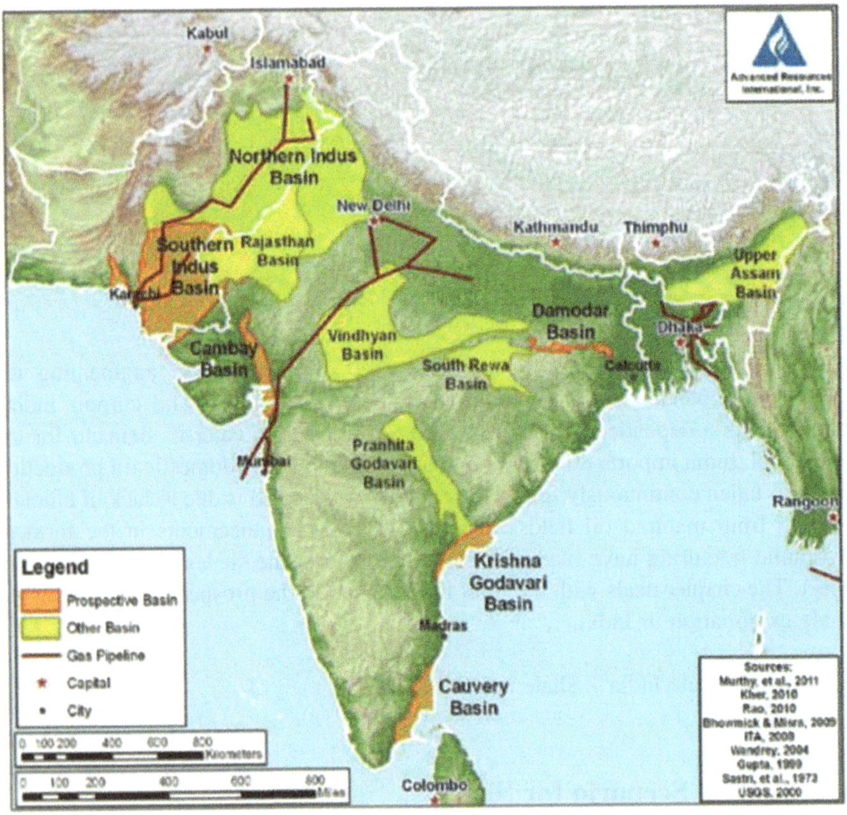

Fig. 7.1 Distribution of identified shale gas basin in India (EIA—2011)

The Indian basins have been reported to geologically complex structure. Cambay basin have been reported to exhibit horst and graben structures. They are also reported to be faulted largely. The distribution of the shale basin in India can be visualized from Fig. 7.1.

7.2 Cambay Basin

It's an elongated basin having intra-cratonic late cretaceous to tertiary rift basin. Its located in the state of Gujarat, India. Basin is predominantly divided into four fault block as mentioned in Table 7.1.

The Cambay Shale is the representation of the marine transgressive that might have occurred in the region. The organic content of the shale is reported to be between Type II and Type III. The total organic content of the shale is reported to be in the range of 2–4%, whereas overall average is reported to be around 2.6%.

Table 7.1 Important fault blocks from Cambay basin

	Fault blocks	Comments
1.	Mehsana-Ahmedabad	Prone to oil
2.	Tarapur	Prone to oil and wet gas
3.	Broach	Prone to oil & wet/dry gas
4.	Narmada	Inadequate data

The thermal maturity is also reported to be in the range of 0.6–2.0% R_o. The prospective depth for this region ranges from 6000 to 16,000 ft. The average oil exploitation zone is reported to be at 8000 ft and the wet gas can be recovered at a depth of about 11,500 ft (Anjirwala 2013; EIA Report 2015; Kumar et al. 2018).

7.3 KG Basin

The basin is located at the eastern coast of the Indian subcontinent spreading over an area of about 7800 m^2. The organic rich shales are reported to be of the Permian to Triassic-age. The shales are extensively good for the oil to dry gas window having its vitrinite reflectance, R_o in the range of 0.7–2.0%. The total organic content is reported to be at max of 11% whereas, the average lies around 6%. The prospective depth for this region ranges from 4000 to 16,400 ft. The average oil exploitation zone is reported to be at a depth 4000 ft. The wet gas can be recovered at a depth of about 8000 ft and the dry gas can be recovered at a depth of about 13,000 ft. The shale resource from the K-G basin is divided into four sub basins (graben) namely Krishna Graben, Gudivada Graben, Bantumilli Graben and Mandapeta Graben (Anjirwala 2013; EIA Report 2015; Kumar et al. 2018).

7.4 Cauvery Basin

The basin is located at the lower eastern coast of the Indian subcontinent spreading over an area of about 9100 m^2. The basin is composed of horsts and grabens, and the rich organic rock lies in the lower Cretaceous Andimadam and Sattapadi Shale. The shales are reported to be of shallow marine Type II to Type III Kerogen. The basin is divided into four major depressions namely Nagapattinam, Tranquebar, Ariyalur-Pondicherry and Thanjavur. The average total organic carbon is reported to be around 2.3%. The basin is found to be rich in thermogenic wet-gas and condensates. The prospective depth for this region ranges from 7000 to 13,000 ft. the wet gas and condensate can be reported be to found at around an average depth of 10,000 ft (Anjirwala 2013; EIA Report 2015).

7.5 Damodar Basin

The Damodar basins is made up of different basins which are collectively referred as the "Gondwanas". It is composed of Satpura, Pranhita-Godavari, Son-Mahanadi and Damodar valley basin. The average total organic content of the coal is found to be about 3.5%, which is favourable for hydrocarbon generation. The thermal maturity of the region is found to be in the range of 1–1.3% R_o. However, the extensive data are not available for thermal maturity and total organic content of the basin. This restricts us to substantiate the different parameters necessary for viable hydrocarbon exploitation.

However, a few other basins like upper Assam basin, an onshore basin in north eastern India. Pranhita-Godavari Basin, a Type III kerogen basin is located in the eastern part of India. Vindhyan Basin, located in the north central part of India is a promising prospect. Rajasthan Basin, a structurally complex basin having Type III Kerogen is located in the north western part of India. The assessment of the thermal maturity and total organic content from these basin is still under developmental stage. Hence they all contribute to probable basins for shale oil and gas exploitation in India (Anjirwala 2013; EIA Report 2015).

References

Anjirwala H, Bhatia M (2013) Shale gas scenario in India and comparison with USA. Int J Sci Res (IJSR) 5:1069–1105

India's hydrocarbon outlook: annual report from the Ministry of Petroleum and Gas. India, 2016–2017

Kumar S, Das S, Bastia R, Ojha K (2018) Mineralogical and morphological characterization of older Cambay shale from North Cambay Basin, India: implication for shale oil/gas development. Mar Petrol Geol https://doi.org/10.1016/j.marpetgeo.2018.07.020

Mitra S (2019) Energizing India: fuelling a billion lives. Rupa Publications. ISBN 978-93-5-333-389-8

U.S. Energy Information Administration (EIA), Report (2015)

Chapter 8
Major Challenges in Shale Exploration

Abstract With the rise of energy-intensive lifestyle, exploitation of Hydrocarbon has been the major focus for the development of nation. Shale oil is the substitute for conventional oil but if the cost is higher than conventional oil, it becomes uneconomical. Shale production is impacted by a number of factors such as political matters, war, social and economic aspect, exploration techniques etc. This chapter summarises the major challenges.

Keywords Shale exploration · Challenges

8.1 Exploration Techniques

Extraction from shales is usually costlier as compared to other conventional resources. Most common exploration technique for shales are hydraulic fracturing and horizontal drilling. In 1940 exploration started by well perforation. Hydraulic fracturing was done commercially for the first time in 1949. In 1965, the technique of hydraulic fracturing was properly discovered (Zendehboudi and Bahadori 2017). In USA, Barnett shales are the first commercially developed reservoirs. The commercial depth for exploration of shales is 1000–5000 m. US has led in exploration of shale gas and oil from reserves.

Fracturing in shales: Most important exploration technology is fracturing. In fracturing, a series of wells are drilled and gas and water and other fluids are forced through the fractures. These fluids along with proppants enter the width of the fractures and push the fractures deeper into the shale bed.

8.2 Challenges in Shale Exploration

(1) In hydraulic fracturing large amount of potable water and toxic chemicals are used which make the drinking water toxic as there is always risk of contamination.

S. Sharma et al., *Unconventional Resources in India: The Way Ahead*,
SpringerBriefs in Petroleum Geoscience & Engineering,
https://doi.org/10.1007/978-3-030-21414-2_8

(2) As there is always pad drilling in hydraulic fracturing operations i.e. wells are drilled in large numbers. Massive use of trucks and emissions (specially in night time drilling operations) put negative effect on the environment too.
(3) Hydraulic fracturing process also affects the communities living nearby.
(4) There may be accidental and occupational health hazard.
(5) In developing countries, there are problems such as governance and corruption and it becomes difficult to provide effective solution to all problems listed above. Also, in these countries the regulatory agencies are not much powerful to stop environmental hazards (Zendehboudi and Bahadori 2017).
(6) Social, economic and cultural challenges.
(7) There might be earth tremors caused by shale gas extraction.

8.3 Government Policy in India

In India, HELP (Hydrocarbon Exploration and Licensing Policy) has created a uniform license for all forms of hydrocarbons-conventional and unconventional oil and gas resources like CBM, shale gas and oil, tight gas and gas hydrates under single license. Before there were separate policy for each of these resources. Open Average Licensing policy has been instituted under HELP to enable fast survey of unexplored geographical areas of the country. Also, E&P companies have been enabled to decide price for the gas and oil produced from high pressure, high temperature reservoirs and deep and ultra—deep water areas. Under HELP, 60,000 crore were committed in October, 2018 for investment in exploration. The awarding was done for 55 oil and gas blocks (Mitra 2019). These oil and gas block amount to 59,000 km^2. Also, national seismic program for unapprised fields was instituted. This program takes a fresh appraisal in all sedimentary basins especially where scanty data is available to have a better understanding of all the sedimentary basins across India to have a better knowledge of hydrocarbon potential. All said and done, still much is needed from policy framework for better exploration of these resources and safety guidelines so that there is least harm to environment.

References

Mitra S (2019) Energizing India: fuelling a billion lives. Rupa Publications. ISBN 978-333-389-8. https://rupapublications.co.in/books/energizing-india-fuelling-a-billion-lives/

Zendehboudi S, Bahadori A (2017) Shale oil and gas handbook, 1st edn. Theory, Technologies, and Challenges, Gulf Professional Publishing

Part II
Shale Conclusion

India is 3rd largest consumer of energy in the world after China and U.S.A. The power of unconventional resources can be defined on the basis that US became the largest producer of oil surpassing Saudi Arabia only on the basis of shale exploration. With the Indian oil Ministry cutting oil imports from November, 2018, it will be difficult for India and shale could be a game changer in current scenario. The main 6 locations (Cambay Basin, Gondwana Basin, Krishna Godavari Basin, Cauvery Basin, Indo-Gangetic Basin and Assam-Arakan Basin) where shale reserves are to be found are populated and fracking (shale exploration technique) may also pollute their ground water resources. Also, fracking can also cause tremors. The way ahead in case of shales would be first to design new and strong legislations regarding shale exploration. Also, indigenization of fracking technology becomes a necessity as exploration will be too costly if we will buy technology from other countries. Keeping these things in mind, shale will be the future of hydrocarbon in India and obviously a way ahead for it.

Part III
Gas Hydrates

Preface

Naturally occurring solid components which are comprised of natural gas and water are known as gas hydrates. They are also known as methane hydrates.

In the last 3–4 decades, there has been a special interest in gas hydrates. There are two driving factors behind this. One is its potential role as an energy resource in the future. Secondly, gas hydrates act as hazards in pipelines and in production facilities and hence needs to be treated.

As far as recoverability of gas hydrates is concerned, major challenge lies in the production of gas from deep marine structures. The gas hydrate sediments should be made accessible by drilling. Next challenge is how to extract methane from methane hydrate. For now, there are three methods available for extraction of methane namely thermal stimulation, depressurization, and addition of chemical inhibitor. Bringing down the cost of producing natural gas from hydrates is yet another challenge.

Worldwide, natural gas hydrates are formed in two main regions, the Arctic and in the ocean. The existence of natural gas hydrates were first confirmed by industry drilling in Arctic permafrost in 1970s. As estimated, the global resource potential of gas hydrates is around 43,300 trillion cubic feet out of which 50% is technically recoverable.

As far as India is concerned, limited available seismic data and preliminary assessment of geological conditions have hinted toward large quantity of gas hydrates in exclusive economic zones of Krishna Godavari (K–G) Basin and Mahanadi Basin. With an estimation of around 933 trillion of cubic feet of natural gas hydrates' deposits, the development of technology to harvest them can give India its energy security and can act as a game-changer.

The purpose of this chapter is to explain what gas hydrates are, under what conditions are they formed, why are they so important, challenges in gas hydrates' production, major players round the world and their role in energy security for India.

Chapter 9
Introduction to Gas Hydrate

Abstract In this chapter "Introduction to Gas hydrate" authors have made an attempt to reveal the truth about the gas hydrate existence and identification of these sources as an energy reserve. The discovery of these vast energy sources is not conventional, during 18 and 19 century these same into existence when science was in its emerging phase and new sources of energy was the need of that time even till today. The history of gas hydrate, their structure and in their existence in the world and especially in India has been briefly discussed.

Keywords Gas hydrate · Introduction to gas hydrate · Structural identification · Locations of natural gas hydrate

9.1 Gas Hydrate

Gas hydrate are the class of clathrate compounds commonly called as "frozen fuels". Clathrate compounds are those chemical compounds where one compound is inhabited inside the other. For this reason, scientists refer these compounds as host-guest complexes. Clathrate hydrates are among the most common natural clathrate compound existing in nature which forms a special type of gas hydrate that consists of water molecules encompassing a trapped gas as a guest molecule. The history of these hydrates exists from back early 1800 where Humphrey Davy, recognized chlorine to be an element. Davy observed a different kind of solid trapping chlorine in ice caged structure. It was in 1823, Michael Faraday, assistant to Humphrey Davy released an official report discussing about the strange ice caged compounds and called it as chlorine clathrate hydrate.

S. Sharma et al., *Unconventional Resources in India: The Way Ahead*,
SpringerBriefs in Petroleum Geoscience & Engineering,
https://doi.org/10.1007/978-3-030-21414-2_9

9.2 History of Gas Hydrate of Hydrogen

In 1930s natural gas miners encountered the problem of ice crystal like material blocking the pipelines when exposed to cold temperature. It was in 1934 when Hammerschmidt through his fundamental knowledge in this field predicted that the freezing or formation of ice structure results in formation of hydrate plug in pipelines and it is not merely ice crystal as identified in the past. He also formulated basic methods to control this hydrate formation which later got advancement in the field of hydrate technology and are exploited till today. In 1940 scientist from Soviet gave a hypothesis regarding the existence of methane hydrate formation in northern regions. In 1945 Prof. Katz gave a diagram predicting the gas hydrate formation pressure and temperature range to implement the gas extraction, transportation process. Later in 1950s a collaboration of D. Katz and R. Kobyayashi developed new methodologies to form a strong base of theoretical studies on gas hydrate. German scientist Von Stackelberg with his team identified the caged compounds at hydrate through X-ray Diffraction technique which were proposed as structure I and II by Claussen in 1951. In late 1950s van der Waals and Platteeuw became the pioneers to identify the condition for hydrate formation. In 1960s Soviets recognized the emerging energy source as gas hydrate and discovered the first major gas hydrate deposit. At this stage, scientist around the world were developing instrumental methods to know more and more about gas hydrate structure. In 1980s using spectroscopic analysis like Nuclear magnetic resonance, Raman spectroscopy Ripmeester et al. gave a new hydrate structure H. In the current scenario, deep research is going on, that emphasize on hydrate formation and its kinetics, production of gas from hydrate reservoirs, transport and storage of gas hydrate. Top laboratories and scientist collaborators from USSR, USA, Canada and other countries have been focused in investigating the methods of survey and prospecting the onshore and offshore natural gas hydrate deposits, and estimating the hydrate resources.

This frozen fuel has an immense potential as energy resource around the world. It is estimated that total amount of methane encapsulated in these hydrate molecule around the world may vary from 120,000 trillion to 320,000 trillion cubic feet. This estimation is very high when compared with the conventional natural gas reserves are expected to be around 13,000 trillion left over on the planet earth.

9.3 Structure of Gas Hydrate

Gas hydrate are the class of compounds that are formed at a specified temperature and pressure conditions. The major forces responsible for the particular type of arrangement in the form of cavities are hydrogen bonding between the water molecules. And the gas molecules acting as the guest between the cavities of the host molecules are linked with van der Waals forces. Due to higher tendency of

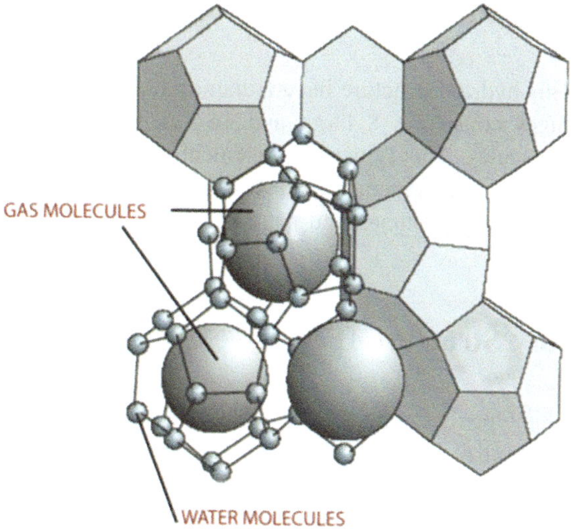

Fig. 9.1 Image depicting the prototype to host-guest hydrate complexes. Adapted from SETARAM

water to form hydrogen bonding in the form a cavity makes it beneficial to entrap the gas molecules inside it without hampering the stability of the gas hydrate at low temperature condition. The most commonly found gas found in these hydrate is methane along with varied amount of different gases like nitrogen, ethane, butane, propane, carbon dioxide, iso-butane and hydrogen sulfide. Figure 9.1 represents basic guest-host complex structure.

During the developing phase of hydrate science several structure for gas hydrate were known among these most common are structure I and structure II proposed by Claussen in 1951. Structure H was discovered by Ripmeester in 1987 through X-ray diffraction analysis of gas hydrate crystal.

The lattice arrangement of hydrate is depicted with reference to water structure. The water molecule exists in the form of a pentagonal dodecahedron arrangement forming the host cavity represented by notations 5^{12} (12 faces having five sides per face). Other arrangements can also be defined as $5^{12}6^2$ (tetrakaidecahedron), $5^{12}6^4$ (hexakaidecahedron), $4^35^66^3$, and $5^{12}6^8$.

9.4 Structure I

This type of structure has a body centered cubic (bcc) arrangement. This hydrate structure is composed of 8 polyhedral cages consisting 6 large and 2 small cages. Hydrate cavity was formed by 46 water molecules forming an arrangement 8G · 46 caged H_2O, where G represents the gas molecule acting as guest. Structure I is usually a small molecule and prominent guest molecule are methane, hydrogen sulphide and carbon dioxide.

9.5 Structure II

This hydrate structure have a diamond lattice type arrangement with 24 polyhedral cages consisting 8 large and 16 small cages which are formed of 136 water molecules. This typical arrangement can be represented as 24 G · 136 H_2O. This type of arrangement is slightly bigger than structure I and the entrapped molecules are generally ethane and natural gas.

9.6 Structure H

This type of structure is a double hydrate structure as it requires types of molecules to stabilize.

Structure H comprises of three types of different cavities where 34 water molecules are linked to three 5^{12} cavity for guest molecules, two $4^{3}5^{6}6^{3}$ cavity for guest molecules and one $5^{12}6^{8}$ cavity for the guest molecules according to the size of the guest molecule.

Basic arrangement of above mentioned structures are shown in Fig. 9.2.

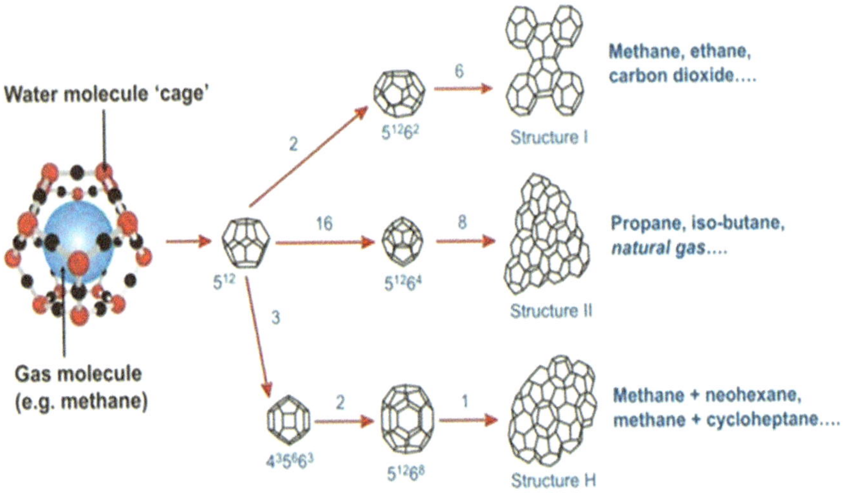

Fig. 9.2 Shows the different arrangement of hydrate cavity in various hydrate structure (Fatnes 2010)

9.7 Hydrates and Their Composition

Hydrate composition can be determined by analyzing the initial parameters of the gas hydrate crystal formed by gas and water under temperature and pressure conditions. It is observed that composition of a single unit hydrate crystal is generally fixed over a varied range of temperature and pressure. During filling of the hydrate cavity molar ratio of the water and gas molecules are expected to change. For naturally occurring gas hydrate composition of the gas, pressure and temperature are responsible for their existence. As the components of the guest molecules will change it may lead to heavy components encapsulating in the hydrate cavity.

9.8 Natural Gas Hydrates

Gas Hydrates in their natural form exist in nature abundantly and are considered as an enormous energy resource left undiscovered in the crust. Natural gas hydrates reserves are expected to exceed the amount of gases present as hydrates may be higher than the total conventional reserve of gases. In the early 1960s, researchers discovered the first natural methane hydrate, or "solid natural gas," existed in the Messoyakha gas field in western Siberia. Scientists like Kvenvolden and Rogers in 2005 identified 89 hydrate sites around the world, out of which 23 locations were able to recover samples and 63 locations referred from Bottom Simulating Reflector (BSR) and 6 remaining locations were interpreted using geological settings

Fig. 9.3 **a** Natural gas hydrate adapted from Zarinabadi (2012). **b** Major methane hydrate fields around the word the world taken from howstuffworks.com

(Trung 2012). Figure 9.3 (a) represents the gas hydrate extracted in the drilling process and (b) represents the major methane hydrate fields.

9.9 Location of Hydrate Zone in India

The National Gas Hydrate Program Expedition (NGHP) started in India focusing on hydrate formations region offshore of the Indian Peninsula and the continental margin in the Andaman. The chief objectives of drilling to these sites was to analyze the geologic and geochemical controls responsible formation of methane gas hydrates and the association with the underlying petro-geology systems. For this purpose, NGHP-01 identified four areas where natural gas hydrate existed: (1) the Kerala-Konkan Basin in the eastern Arabian Sea, (2) the Mahanadi basin (3) Krishna-Godavari Basins in the western Bay of Bengal, and (4) the Andaman fore-arc Basin in the Andaman Sea (Lorenson and Collett Lorenson and Collett 2018) (Fig. 9.4).

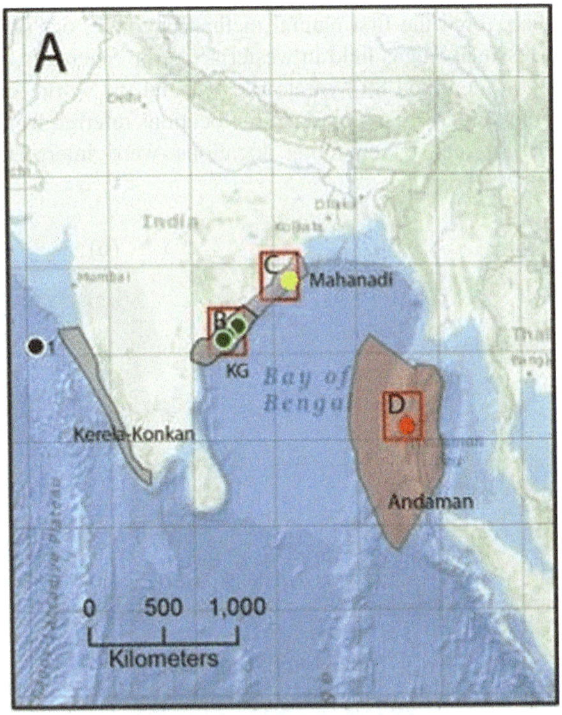

Fig. 9.4 Major gas hydrate sites identified in India (Lorenson and Collett 2018)

References

Fatnes ED (2010) Numerical simulations of the flow and plugging behaviour of hydrate particles. Master's thesis. The University of Bergen
Lorenson TD, Collett TS (2018) National gas hydrate program expedition 01 offshore India; gas hydrate systems as revealed by hydrocarbon gas geochemistry. Mar Pet Geol 92:477–492
Trung NN (2012) The gas hydrate potential in the South China Sea. J Petrol Sci Eng 88:41–47
Zarinabadi S, Samimi A (2011) Problems of hydrate formation in oil and gas pipes deals. Austr J Basic Appl Sci 5(12):741–745

Chapter 10
Hydrates and Their Properties

Abstract Even after the known history of almost 200 years, hydrates and their properties were difficult to study due to their complexity in the research. The properties of the hydrate system were determined by their structure and the characteristic composition of the hydrate cluster. In the present day over 100 types of clathrate structure are known which exist as a multicomponent system as their size is so large it makes large clusters. The new technologies like X-ray diffraction, Nuclear magnetic resonance (NMR) have made it easier to study the hydrate structure and its properties in depth.

Keywords Gas hydrate properties · Thermal stability of hydrates

10.1 Thermal Properties of Gas Hydrates

Thermo-physical and thermochemical properties of a gas hydrate system are very important in all the natural and techno-genic hydrates explored during the course of study. Due to complexity in the hydrate structure the properties predicted were poorly studies and showed negligible consistency in the results so it was very difficult to form a generalized idea about the properties studied. Though instrumental results were considered as the most reliable method to determine the thermal properties of the gas hydrate system and mostly analytical techniques were used to determine heat of formation of gas hydrate, heat capacity and thermal conductivity.

It was in 1979 Stoll and Brayan were successful in measuring the thermal conductivity of and found that the propane hydrates have much lower thermal conductivities of ice (Stoll and Brayan 1979). Solid hydrates also showed the similar trend as that of liquid water and had thermal conductivities around (0.605 Wm^{-1} K^{-1}) (Fig. 10.1).

In 1981 Ross et al. found the THF hydrate showed no pressure dependence and their thermal properties were based on temperature dependence (Ross et al. 1981). Generally at laboratory scale the thermal conductivities of THF hydrate system are studies as these hydrates are miscible in water. In 2005 Waite et al. showed that

S. Sharma et al., *Unconventional Resources in India: The Way Ahead*,
SpringerBriefs in Petroleum Geoscience & Engineering,
https://doi.org/10.1007/978-3-030-21414-2_10

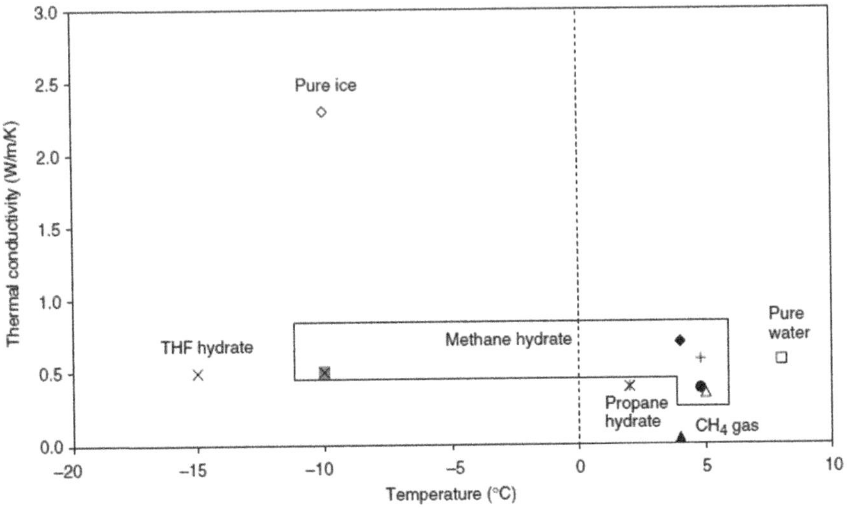

10.1 Thermal conductivities of gas, water, ice and hydrated without sediments system. Adapted from Gupta (2007)

THF hydrates have similar thermal conductivities as that of methane hydrates and its value lies around 0.50 Wm^{-1} K^{-1} (Waite 2005).

10.2 Electrical Conductivity of Gas Hydrates

Electrical properties of gas hydrate system are important to determine as during hydration formation studies the control is monitored using electric fields of various frequency and strength. These electrical testing signals penetrate the hydrate saturated rocks where these studies are carried out. Makogon in 1974, was pioneer in studying the electrical resistance of the gas hydrate system in real cores (Makogon 1974). Later in 1980s electrical resistivity's of hydrates were studied on THF hydrates (Cox 1983). Electrical resistance has not been employed to determine the hydrate content in layers during the well testings.

10.3 Density

Among all the properties of hydrates under study density is chief characteristics of the gas hydrates as it is totally govern by the hydrate structure and the type of guest inhabiting the host complex. Hydrate density measurement was a tedious task for almost a century, it was possible only by using NMR and EPR technologies to measure hydrate density wen the hydrate structure is known as their lattice constants are generally known. it was found that the average gas molecule distances in a hydrate structure I is approx. 6 Å and that of gas molecule is 33.4 Å and that of water is 3.1 Å. These value clearly indicates that the physical properties of gas hydrates ice between pure ice and pure water.

References

Cox JL (1983) Natural gas hydrates: properties, occurrence and recovery. United States

Gupta A (2007) Methane hydrate dissociation measurements and modeling: the role of heat transfer and reaction kinetics. Ph.D. thesis. Colorado School of Mines, Golden, CO

Makogon YF (1974) Hydrate of natural gas. NEDRA, Moscow

Ross RG, Andersson P, Backstrom G (1981) Unusual PT dependence of thermal conductivity for a clathrate hydrate. Nature 290:322

Stoll RD, Bryan GM (1979) Physical properties of sediments containing gas hydrates. J Geophys Res 84:1629–1634. https://doi.org/10.1029/JB084iB04p01629

Waite WF, Gilbert LY, Winters WJ, Mason DH (2005) In: Proceedings of the fifth international conference on gas hydrates. Paper 5042. Trondheim, Norway, 13–16 June 2005

Chapter 11
Gas Hydrate Formation

Abstract Gas hydrates are solids that resembles ice when natural gas and free underground water combined to form hydrate structure at high temperature and low pressure condition. Gas hydrate formation can take place ain gas or condensed wells and also in deep oil wells rich in natural gas. Formation of hydrates in a well depends on various factors like composition of the fluid, design and operating regimes, geothermal gradient in the well. Gas hydrates are rich source of unconventional energies to the world and very few knowledge about them has been gathered until now and researchers around the world are active to know more and more about these gas hydrate systems.

Keywords Hydrate formation · Hydrate problems

11.1 Experimental Condition for Gas Hydrate Formation

The researchers around the world has been curious to know about the structure and the energies associated in the bonding of this multicomponent system. There is need to develop a theory about the development of the hydrate crystal and its hydrate properties. Various mechanism and techniques has been proposed to study the parameters associated with the hydrate crystal formation. Scientist have proposed two procedures to acquire the formation condition of the gas hydrate. Out of these proposed mechanism the first one is the "temperature search" method in which formation temperature is analysed by keeping the pressure of the system constant. The another method is the "pressure search" mechanism where formation pressure is obtained by keeping the temperature constant (Mei 1996). The researchers have developed the apparatus known as formation cells which are used as a prototype of the wells where has hydrate are formed in nature. These cells have an operating pressure up to 2200 MPa where the hydrate formation of different gas and their mixtures where studied, and it was predicted that high pressure results in hydrate formation (Zhang 2006).

S. Sharma et al., *Unconventional Resources in India: The Way Ahead*,
SpringerBriefs in Petroleum Geoscience & Engineering,
https://doi.org/10.1007/978-3-030-21414-2_11

11.2 Hydrate Formation in Wells

Hydrate formation can take place where free gas, water and suitable pressure and temperature condition occurs. Gas hydrate also occur when the liquid temperature reaches below the equilibrium temperature due to large depressions, or when the pumping water temperature reaches below the hydrate equilibrium in the bottom well zone or the oil-gas zone. Hydrate formation occurs in the production pipes column especially in the throttle device. It was once observed in the offshore fields, when the wells are shutdown the hydrate formation chokes the wellbore, or even can damage the well construction. Hydrate formation can take place in newly drilled well if is uses water solution as low temperature and at high pressure conditions and in such condition hydrate formation increase the cost of the production. Formation of hydrates in a well depends on various factors like composition of the fluid, design and operating regimes, geothermal gradient in the well. The mechanism that results in formation of the gas hydrate in the shut in oil well, which generates the solid slurry and choke the pipe line. The reason behind this hydrate formation is that under HP/HT condition oil is produced from the wellbore, the liquid water arises from the solution that results in lowering of the temperature and forms a suspension in the dorm of microdroplets. These droplets aggregate and precipitates. The water is now saturated with the gas which results in formation of hydrate during suitable pressure temperature condition.

11.3 Pressure–Temperature Condition for Gas Hydrate Formation

There are five perquisite conditions for the formation of the gas hydrate: presence of free water, low temperature, high operating pressure, high velocities or pressure pulsations and presence of hydrocarbon and sulphur source. Gas hydrate system is a gas-solution system where the two constituents are not actually chemically bounded, their co-existence in the wells and geothermal locations is only due to the formation of appropriate temperature–pressure condition in that reason that supports the gas hydrate formation Fig. 11.1 represents the phase diagram of the water and hydrocarbon system existing together and relates the formation of hydrate crystal.

The image clearly indicates that the formation of gas hydrate system is in high pressure and low temperature region. The C point in the figure is the critical point where hydrocarbon and water phase exists in the equilibrium. Q1 and Q2 is the quadruple point where the four system exists in equilibrium. During deep water dwelling, the gas hydrates are found to be stable in low temperature zones and as the temperature increase the dissociation of the hydrate occurs as explained in Fig. 11.2 (Alcázar-Vara 2018).

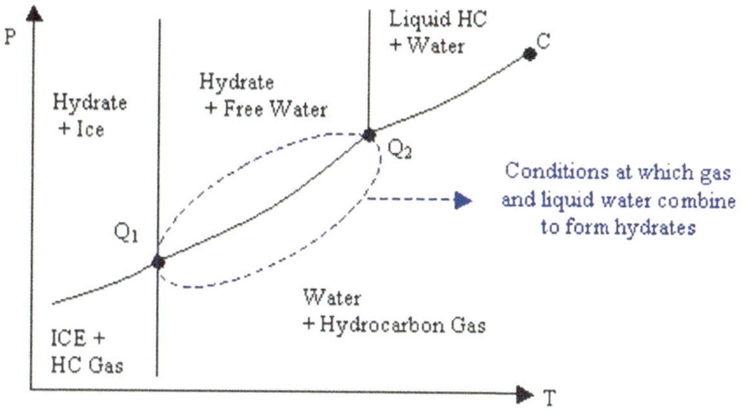

Fig. 11.1 Phase behavior of water and pure hydrocarbon system. *Source* Figure taken from courseware module of Penn State's College of Earth and Mineral Sciences

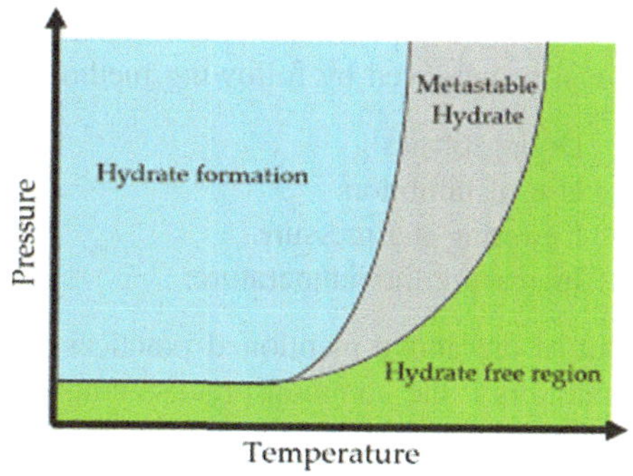

Fig. 11.2 Represents the typical phase diagram of hydrate stability zones (Alcázar-Vara 2018)

11.4 Kinetics of Gas Hydrate Formation

Previous studies have predicted that gas hydrates forms the potential resources of natural gas and is expected to be distributed all over the land and sea. To predict the possibility of producing natural gas from the gas hydrates it is very important to analyse each and every parameter associated with the gas hydrates. Out of these our major concerns are: reservoir and their geologic arrangements, structure of hydrate in the porous medium, its permeability, energy required to dissociate the hydrate and finally the kinetics of its formation. Around the world many articles have been published about the formation of gas hydrate in dynamic conditions. They have obtained hydrate by subsurface gas-aqueous contact by mixing or bubbling.

The kinetic study of the hydrate formation was only viewed to study hydrate accumulation and supercooling of the fluid. Since these gas hydrate crystal serve as an important system for transportation of the gas and desalination of water, it became mandatory to know more about these crystals. Scientist evaluated that the nucleation of hydrate crystal chiefly methane, ethane was an interfacial phenomenon it indicated that this was mainly govern my temperature, pressure and degree of suprecooling (Vysniauskas 1983). To control the formation of these crystal is being important to analyse the appropriate temperature–pressure condition. For these purpose several PVT simulators were used, that assembles the study of activity of water, effect of hydrocarbons and their molecular weight and hydrate inhibitor to simulate a proper system that is efficient in hydrate formation.

11.5 Inhibition of Gas Hydrate Formation Process

Hydrate crystal cause problem in oil and gas industry as they choke the well bore during drilling process, transportation of oil is effected due to formation of gas hydrate in the pipelines. So there is need to restrict the formation of gas hydrate and it can be achieved by following methods:

- Drying of gas
- Use of inhibitors
- Lowering the pressure
- Increasing the temperature.

Out of the above mentioned practices to restrict the formation of hydrate use of inhibitors it the common practise employed in the oil and gas industry. Inhibitors used can bee of two types: thermodynamic and kinetic. In thermodynamic restriction by the use of inhibitor as components is added to the water gas system that disrupt the equilibrium of the system and hence no formation of gas hydrate occurs. Alcohols and salts are the common thermodynamic inhibitors used as when added to the system they alters the hydrogen binding capacity of the water molecules which alters the thermodynamic equilibrium existing between the water gas systems.

The kinetic inhibitors get itself adsorbed on the surface of hydrate crystals and disrupts the diffusional exchange at the water-inhibitor-gas interface, decrease the formation of hydrate crystals by coagulation with the water molecules and later cause sedimentation of the crystals. Kinetic inhibitors do not interfere with the formation chemistry but it shifts the time to attain the equilibrium. These inhibitors are soluble in water and gets adhere with the polar surface of the hydrate crystal and creating a barrier between the other hydrate crystal to form a cage like structure. Fatty acids and poly-*N*-vinylpyrolidone (PVP) is commonly use kinetic inhibitor to avoid the formation of gas hydrates.

References

Alcázar-Vara LA, Cortés-Monroy IR (2018) Drilling fluids for deepwater fields: an overview. In: Recent insights in petroleum science and engineering. InTech

Mei D-H, Liao J, Yang J-T, Guo T-M (1996) Experimental and modeling studies on the hydrate formation of a methane + nitrogen gas mixture in the presence of aqueous electrolyte solutions. Ind Eng Chem Res 35(11):4342–4347

Vysniauskas A, Bishnoi PR (1983) A kinetic study of methane hydrate formation. Chem Eng Sci 38(7):1061–1072

Zhang L-W, Qiang H, Sun C-Y, Ma Q-L, Chen G-J (2006) Hydrate formation conditions of methane + ethylene + tetrahydrofuran + water systems. J Chem Eng Data 51(2):419–422

Chapter 12
Application of Gas Hydrates

Abstract The existence of natural gas hydrate system in the earth crust is both a boon and a bane to the technological advancement in the oil and gas industry. At one stage the properties, parameters associated with the hydrate formation, and their characteristics to exist in the hydrate structure formed by transition of free bound water molecules at high pressure and low temperature condition pave the way to new technological advancement. At another stage these hydrate create problem when they choke the oil and gas flow lines. But these hydrates system is among those unconventional resource of energy that is present in huge deposits in permafrost regions. It is roughly estimated that the total natural gas reserves in these gas hydrate is almost the double of all the fossil fuel reserves available around the world.

Kewords Gas hydrate application · Storage and transportation in Gas Hydrate

12.1 Storage and Transportation of Gases in Gas Hydrate Clathrate

Gas hydrate clathrate are the cage like structure that has an unique property of changing its specific volume during its transition from free water stage to bounded clathrate type structure. During the transition stage of water to ice it is expected that approximately 26–30% of the volume space is increased. During the research it has been observed that the volume space of the hydrate crystal formed may change by several folds in magnitude. This interesting property of the hydrate crystal has been employed in transportation and storage of large volumes of gases at low temperature high pressure condition only as this in mandatory condition for the existence of stable hydrate structure. Natural gas needs to be transported from the reserve area to the areas of high demand and it needs to be stored till demands exceeds the production. It is reported that 1 ft^3 of the hydrate crystal can hold upto 150–180 ft^3 of natural gas approximately (Pooladi-Darvish 2004). Natural gas is the cleanest and the cheapest source of domestic fuel obtained from the fossil fuel. In this process the water molecules are bonded in the form of a soccer ball and the gas molecule of an appropriate size fits itself in the cavity created by the bonded water molecules at low

S. Sharma et al., *Unconventional Resources in India: The Way Ahead*,
SpringerBriefs in Petroleum Geoscience & Engineering,
https://doi.org/10.1007/978-3-030-21414-2_12

temperature and high pressure condition. This 3-D structure encloses the gas molecule in the clathrate structure having a gas density near to the liquefied natural gas (LNG). A group of scientist in Iran conducted the research on production, transportation and re-gasification and the cost estimation of the natural gas hydrate (NGH) source of energy to that with the LNG (Taheri 2014). The hydrates in transport of gases has been studied mainly for the natural gas and carbon dioxide transportation. Hydrates can be used in creating a compressor less system for increasing the pressure of gas, production of energy by employing low-potential heat.

12.2 Desalinization of Aqueous Solutions

Out of the 1.5 billion cubic kilometres of the water existing on the planet earth, 97% is constituted to the saline water that comprises of the sea and oceans. The remaining 3% of the fresh water for the drinking purpose constitute to approx. 30 million kilometres, out of this most of the water reserves are in polar and mountain glaciers and are beyond the reach of humans. Only 3% of the 30 million km^3 is available to humans and is found in lakes and rivers which is employed in agricultural and other human practices. This natural water reserves are unevenly distributed across the globe and the fresh water resources are concentrated to large rivers like Ontario, Michigan, Baikal etc. Artificial reservoirs of water on river banks has been set up to desalinize the ground water and to use it in the irrigation process. Desalination of water using hydrates has been brought into practise because of following properties of the hydrate.

1. Hydrate molecule is a pure bond formed by the hydrogen bonding between the molecules at very low temperature and high pressure condition, agents remove the inorganic and organic impurity in the solution, which lead to formation of pure water hydrate.
2. The difference in the crystal densities of two separate phases allow them to form hydrate and salt crystal separately.

In the desalination process the hydrate crystal is first de-aerated, transformed into solid at low temperature. This transformation into hydrate crystal takes place at slight lower temperature then the freezing points due to the colligative property of depression in freezing point due to the presence of electrolytes in the form of salts. The phase density difference of the crystal is utilized to separate the crystal which are melted and this de-saline water is used as a source of fresh water.

12.3 Hydrate Plug in Well Bore

Underground blowouts are the most common problem involving the flow of fluid from high pressure area to low pressure areas. The blowouts in the production or the drilling wells ate the most common ones arising due to tubular corrosions. Very less number of statics and reports have been published about these as in most cases the actual cause is rarely identified. One such blowout happened in 1981 on the Pechora River on the shore of Arctic Ocean (Makogon 1997). During the research about the hydrates it was found that these crystals have good strength and adhesive properties which can be employed to curb the uncontrollable gases and fluid flowing through the wells during the blowout of wells. This method of creating a hydrate plug in the production or drilling well to curb the blowout is an cost effective method and less complicated than other methods like drilling a parallel well, drilling well into the emergency well and using rock shift method to kill the blow out in the well. To create a plug in the well water at its freezing point is injected with high pressure that results in the formation of the hydrate plug which clogs the well and stops the release of the harmful gas.

12.4 Concentrating Heavy Water

The presence of natural gas hydrate contribute a lot to sustainable economic development. Properties of hydrate make them as a chief role player in the field of energy reserve and the maintenance of the pollution free environment. Formation of the hydrates results in the exclusion of the salt from the system, and for this reason these hydrates have found their application in removal of the metallic impurities from the water system. The contamination of the water source with heavy metals like Cd (II), Hg (II), Ni (II) and Cu (II), adversely affect the environment and cause serious health effects like kidney dysfunction and hormonal imbalance to the human population inhabiting near the contaminated water body and consuming it. Several industries where effluent contains heavy metals directly discharged in the water body without any effluent treatment pollutes the water resources. To remove these heavy metals from the system several methods like ion exchange, adsorption, membrane filtration etc. Though these method are effective in removing the metals but are not cost effective and have some special requirements like pH and temperature, which restricts their application for the bulk process. For this process and alternate method of contaminated water treatment was studied which was based on hydrate separation technique.

To develop this technique of water treatment using hydrates scientists have developed a system that contains five main steps: hydrate formation, vacuum filtration, centrifugation, dissociation of hydrate and the analysis of water after treatment (Dong 2017). The removal efficiency of the waste water was calculated after final washing, that clearly predicts approximately the concentration of the heavy metal ions decreased by 20–30%. In this process of treatment the metals ions which earlier was partially trapped in the hydrate cage were dissociated in the water. The treated water is now washed to remove these heavy metal ions (Song 2016).

12.5 Carbon Dioxide Sequestration

The enormous availability of natural gas in natural gas hydrates gives us an opportunity to employ it in CO_2 capture and its storage as a future source of energy. From the global warning security issue there is need to replace the methane gas with carbon dioxide in the natural gas hydrate reservoir. This method of removing methane and paving the clean and efficient process of CO_2 capture mitigates the global warning and reduces the geo-mechanical hazards caused by using conventional methods (Lee 2015). CO_2 capture is an energy consuming continuous process. The recent technologies involving CO_2 capture are absorption, adsorption and membrane utilization, but these process are less efficient and has high operating cost. The new emerging technology for CO_2 capture is the hydrate base CO_2 capture which is in discussion from past two decades both in industry and academia. This process operates at mild temperature and pressure condition, where the specific mechanism of separation is employed for separating the gaseous mixture. The pioneer studies in this area involves the studies of phase equilibrium analysis of CO_2 capture, but later in the recent advancements it deals with mixing of chemical additives and mechanical techniques for efficient CO_2 capture (Dashti 2015).

It has been found in the research that small molecules of CO_2 gas from a structure I (S_1) in the hydrate crystal. The CO_2 molecule is non-polar hydrocarbon which forms an S_1 hydrate crystal when brought in contact at low equilibrium temperature and pressure condition. This crystal on dissociation at STP conditions releases approx. 170 volume of CO_2 gas (Dyadin 1999). Figure 12.1 the curve depict that CO_2 has low hydrate forming temperature and pressure condition when compared with the other gas live nitrogen, hydrogen and oxygen.

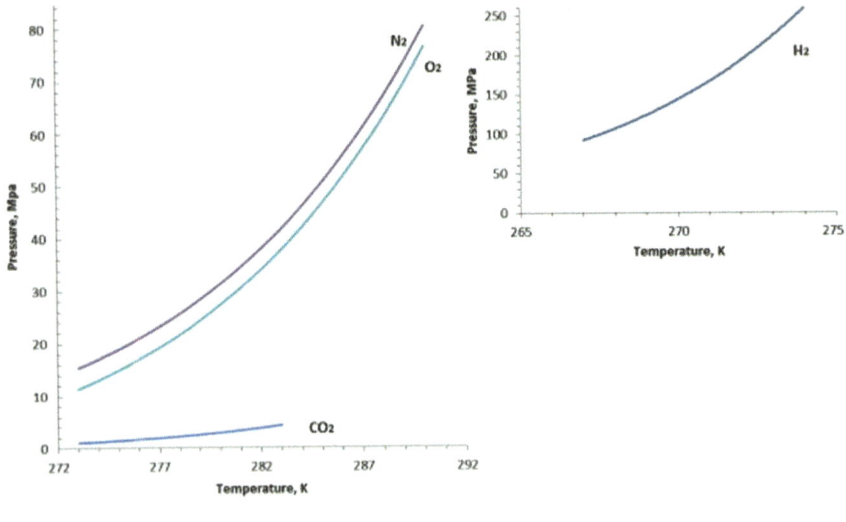

Fig. 12.1 Equilibrium phase diagrams for different hydrate formers (Dyadin 1999)

Figure 12.1 the curve depict that CO_2 has low hydrate forming temperature and pressure condition when compared with the other gas live nitrogen, hydrogen and oxygen. The dissociation of the gas after hydrate formation is much higher than the initial gas concentration as a feed. For this reason the process of Carbon dioxide sequestration is an energy efficient and environmentally friendly process.

References

Dashti H, Yew LZ, Lou X (2015) Recent advances in gas hydrate-based CO_2 capture. J Nat Gas Sci Eng 23:195–207

Dong H, Fan Z, Wang B, Xue S, Zhao J, Song Y (2017) Hydrate-based reduction of heavy metal ion from aqueous solution. Energy Proc 105:4706–4712

Dyadin YA, Larionov EG, Manakov AY, Zhurko FV, Aladko EY, Mikina TV, Komarov VY (1999) Clathrate hydrates of hydrogen and neon. Mendeleev Commun 9(5):209–210

Lee Y, Kim Y, Lee J, Lee H, Seo Y (2015) CH_4 recovery and CO_2 sequestration using flue gas in natural gas hydrates as revealed by a micro-differential scanning calorimeter. Appl Energy 150:120–127

Makogon IF, Makogon IUF, Makogon YF (1997) Hydrates of hydrocarbons. Pennwell Books

Pooladi-Darvish M (2004) Gas production from hydrate reservoirs and its modeling. J Pet Technol 56(6):65–71

Song Y, Dong H, Yang L, Yang M, Li Y, Ling Z, Zhao J (2016) Hydrate-based heavy metal separation from aqueous solution. Sci Rep 6:21389

Taheri Z, Shabani MR, Nazari K, Mehdizaheh A (2014) Natural gas transportation and storage by hydrate technology: Iran case study. J Nat Gas Sci Eng 21:846–849

Chapter 13
Challenges in Gas Hydrate Formation in Oil Industry

Abstract Gas hydrates hydrocarbon system though emerging as a leading source of energy sin the form of natural gas but the formation of hydrates in drilling processes creates problem and subjecting to the production loses and increasing the cost of operating process as high amount of capital goes wasted in inhibiting their formation or removing the hydrates crystals. A major problem arose for the oil and gas industry to develop a research technique that is capable in reducing the risks caused by the hydrate formations. Common problems associated with the hydrates plugs in wells which is a very tedious task to remove these hydrates. It is almost impossible to remove these hydrates, so to inhibit the formation during the production is the only way out, addition of inhibitors, gas lift are the methods to retard the growth of hydrate formation.

Keywords Hydrate problems · Flow assurance · Hydrate inhibitors · Drilling problems

13.1 Occurrence of Gas Hydrate in Drilling Processes in Offshore Regions

In India, Ministry of Petroleum & Natural Gas under National Gas Hydrate Program (NGHP) has been dealing with the research, exploration and production activities from the gas hydrates. The presence of Gas Hydrate has been reported in Krishna Godavari basin, Mahanadi region, Gulf of Mannar and Andaman Basin. It has been reported that a total of 39 wells has been on 21 locations for the conformation of the existence of natural gas hydrates, Majority of the location have been offshore in clay dominated settings (DGH report 2016–2017).

The new challenges being faced during the commercial exploitation of hydrocarbon from these fields. The problem during drilling are analogous to the problems faced by the rest of the industry worldwide. The major problems associated with the drilling in hydrate prone zone are of:

S. Sharma et al., *Unconventional Resources in India: The Way Ahead*,
SpringerBriefs in Petroleum Geoscience & Engineering,
https://doi.org/10.1007/978-3-030-21414-2_13

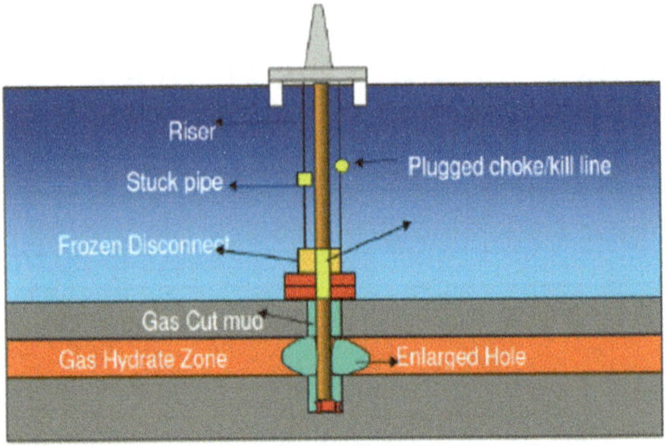

Fig. 13.1 Representation of the problems encountered while drilling through NGH (Amodu 2008)

(1) Borehole instability: It has been observed that majority of the sediments are consolidated when the gas hydrates are associated with the clay matrix. The hydrate structure provides the required degree of cementation to keep the borehole consolidated. During drilling process, the hydrate stability may be lost which can reduce the consolidation thereby, leading to cave-ins.
(2) The dissociated hydrates are found to alter the properties of the drilling fluid being used making it susceptible to failure in bottom hole condition.
(3) The enormous amount of gas which is liberated during the dissociation of NGH can lead the well susceptible to kick or a blowout.

Further, more challenges are encountered during offshore drilling for oil and gas at increased depths. It has been found that when the shallow sediments containing natural gas are drilled, the gas enters into the drilling fluid and can form gas hydrates. This is attributed to the enhanced pressure and reduced temperature conditions while drilling offshore. If the inhibitors are not added into the drilling fluid under consideration the gas hydrates can be formed when the gases move into the drilling fluid. The gas hydrates can be formed and can block the pipe, annular spaces or it can prevent the closing/opening of the valves of the blow out preventer stack (Amodu 2008; Ning et al. 2010) The gas hydrates can also block the chokes and kill line which can lead to catastrophic conditions. Major problems associated with the drilling of NGH forming zone can be illustrated in Fig. 13.1.

13.2 Flow Assurance Issues

The problem of the flow assurance is among the major problems associated with the offshore developments. This has been listed by 110 energy companies all over the world. The problems can be arranged in descending order of their significance as hydrates, petroleum wax, scales, corrosion and asphaltenes.

Hydrates form at a rapid rate in both onshore and offshore applications. This can plug and disrupt the system flows. The hydrates can form a solid slurry during the flow from well during production or transporting the produced crude to the desired location. They can also be formed at the walls restricting the flow area. The formed hydrate separates the pipes into two distinct zones of higher and lower pressure. The high pressure zone is formed between the source (gas inlet point). The low pressure zone is encountered between the plug and outlet. The high pressure can lead to the bursting of pipe making the operation hazardous (Jassim et al. 2010; Borden 2014).

To combat the problem of hydrate formation and to make sure that the continuous flow is maintained, following precautionary measures have been suggested

(1) Addition of chemical additives
(2) Hydraulic
(3) Thermal methods.

The chemical additives are normally termed as thermodynamic inhibitors. They are designed to inhibit the formation of hydrates by altering the natural gas hydrate stability zone (NGHSZ). The inhibitors move the NGHSZ to a further low temperature and high pressure condition as compared to the original conditions. However, it has been reported numerous time that more than 50% by mass of methanol and glycol is needed to suppress the hydrate formation. This makes the process environmentally and economically unviable. The researchers have developed polymer based inhibitor that further reduced the dosage required for the prevention of propagation of hydrates (Yang and Tohidi 2011). These chemical have been sub categorized as low dosage kinetic inhibitor (LDKI) and anti-agglomerates (AA). Both of these chemicals can't stop the hydrate formations, but just reduce or delay the formation of hydrates when introduce in concentrations as low as 1 mass%.

The hydraulic method deals with the depressurization of gas hydrates leading to the its dissociation. The methods hold good for gaseous hydrates but it can't address the hydrates formed by liquid hydrocarbons. The depressurization of liquid hydrates leads to vaporization which creates another problem. The thermal methods aim at delivering the heat to the plug through the walls of the pipe. This delivers the heat to the hydrate plug initiating its dissociation. However, this method fails to deliver when subjected to treatment of problems in subsea equipment's. The hydrate blocked pipe can also be cleaned by mechanically pigging the system. The pigging involves shooting of a dart (equal to inside diameter of pipe) made up of a flexible polymer. The dart is retrieved at the other end of the pipe section under consideration. However, its uneconomical to stop the operation and performing pipping operation (Sami et al. 2013).

References

Amodu AA (2008) Drilling through gas hydrate formations: possible problems and suggested solutions. Thesis. Texas A&M University

Borden K (2014) Flow assurance, hydrate and paraffin management. Oil Gas Facil 3(1)

India's hydrocarbon outlook: annual report from the Ministry of Petroleum and Gas, India (2016–2017)

Jassim E, Abdi MA, Muzychka Y (2010) A new approach to investigate hydrate deposition in gas-dominated flowline. J Nat Gas Sci Eng 2:163–177

Ning F, Zhang L, Tu Y, Jiang G, Shi M (2010) Gas-hydrate formation, agglomeration and inhibition in oil-based drilling fluids for deep-water drilling. J Nat Gas Chem 19:234–240

Sami NA, Sangwai JS, Balasubramanian N (2013) Gas hydrate applications and problems in oil and gas industry. Int J Sci Eng Res 4(8)

Yang J, Tohidi B (2011) Characterization of inhibition mechanisms of kinetic hydrate inhibitors using ultrasonic test technique. Chem Eng Sci 66:278–283

Chapter 14
Gas Hydrate Scenario in India

Abstract In this present chapter "Gas hydrate scenario in India" an effort has been put together to study the clathrate hydrate system in India. The pioneering agency from India and US collaborated and discovered the hydrate reserves in India. Various expedition from 2006 to 2015 has been conducted to through study the exact location of these reserves in deep sea waters. New technologies has been identified to drill these reserves and attempts are going to avail these energy resources to the common man.

Keywords Indian gas hydrate · Locations · K-G basin · NGHP

Methane gas hydrate system is estimated to provide enormously large volumes of methane gas that are encaged within a hydrate, this large reserve of unconventional energy is the pioneer driving force for the growing interest of the researchers in the field of gas hydrates. In the past few decades the studies have been focused in estimating the volumetric reserve of hydrates in petroleum geology, that governs the occurrence of natural hydrates. Through studies it has been found that the transition zone among the hydrates and natural gas is called as the "base of the gas hydrate stability zone (BGHSZ)". The occurrence of hydrates in the sedimentary rocks enhances the velocity whereas existence of the free gas lowers the velocity, hence creasing the stability zone. The BSR (Bottom simulating reflectors) are assumed to be the best indicators of hydrate presence in nature. Some geophysical and geochemical proxies has been identified in KG offshore, Mahanadi, and Andaman basins, India (Ramana et al. 2006; Dewangan et al. 2011).

The NGHP-01 expedition was executed under the various Agencies mainly "Directorate General of Hydrocarbons (DGH) under Ministry of Petroleum and Natural Gas (India), U.S. Geological Survey (USGS), Consortium for Scientific Methane Hydrate Investigations (CSMHI) under Overseas Drilling Limited (ODL) and FUGRO McClelland Marine Geosciences (FUGRO)" (Collett et al. 2008). With the tremendous efforts of Indian National Gas Hydrate Program (NGHP) Expedition 01 in 2006 the ship JOIDES Resolution drilled approximately 39 holes at 21 different sides and found reserves of gas hydrates at 1 one site in

S. Sharma et al., *Unconventional Resources in India: The Way Ahead*,
SpringerBriefs in Petroleum Geoscience & Engineering,
https://doi.org/10.1007/978-3-030-21414-2_14

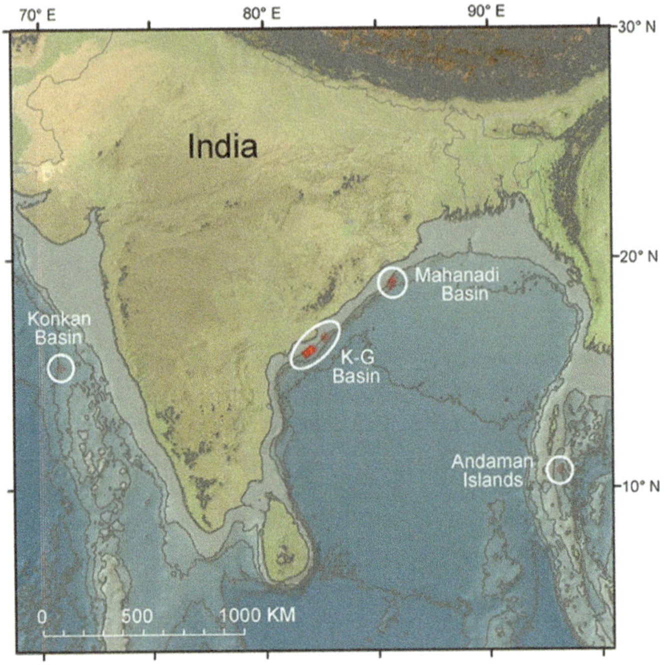

Fig. 14.1 Gas reserves identified in the Indian Ocean. *Source* https://www.maritime-executive.com/article/gas-hydrate-found-in-indian-ocean

Kerala–Konkan basin, 15 new sites in KG basin, 4 sites in Mahandi basin and one site in Andaman regions (Shankar et al. 2013) as shown in Fig. 14.1. In this expedition coring, wireline logging (WL) and logging-while-drilling (LWD) methods was conducted and among all the sites the most valuable datasets were obtained at Site NGHP-01-10. Later in 2007 a scientific drilling of gas hydrates near Indian continental margins was performed which was the joint venture between DGH and USGS.

During NGHP-01-05 and 07 report it was predicted that the Krishna–Godavari Basin mainly composed of the clay stratigraphy and the hydrate crystals of gas occurred in the facture at study sites. This study also predicted that the gas hydrate can also be formed in shallow clay-dominated marine sediments. The convention hydrocarbon system present in K-G Basin reveals that it is the remains of organic matter from the Paleocene and Cretaceous time that resulted in conventional gas and gas-condensate reserves in K-G Basin (Banerjie et al. 1994). The methane gas hydrate crystals obtained during NGHP-01expedition where majorly derived from microbial origin (Collett et al. 2008). Scientists in India were very much instrumental with the research after NGHP-01 as the hydrates discovered were not being extracted as they existed in fractured shale and clay sediments. From 2007 to 2014 a wide study was conducted to recognize the existence of gas hydrates in Eastern offshore which were majorly sand dominated.

In March, 2015 NGHP-Expedition-02 commonly called NGHP-02 was started aiming the exploration oh hydrates in reserves dominated by sand, and to analyze the geological control on hydrate morphology and sediment system. In the eastern offshore in K-G basin the NGHP-02 developed an efficient petroleum based gas hydrate system and discovered a large depositional system which was interconnected in the reservoir. NGHP-02 was successfully completed in July, 2015 with the discovery of thickest gas hydrate sand bearing reservoir (Shukla et al. 2018; Kumar et al. 2018). In Fig. 14.2, the purple dots are the sites drilled during NGHP-01 whereas yellow dots depicts the site drilled during NGHP-02.

During NGHP-02, gas hydrate petroleum system (GHPS) approach was used for understanding the occurrence of hydrate and sediment property by four components:

1. Temperature and pressure condition near hydrate stability zone.
2. Methane and water supply within sediments.
3. Reservoir sediments to hold the gas hydrate.
4. Proper sealing to avoid the leakage of gas trapped within the reservoir.

With the discovery of the potential gas hydrate reserves in the eastern offshore the government of India has been very much interested to avail these unconventional resource for the benefits of the common man. The industrial software's were

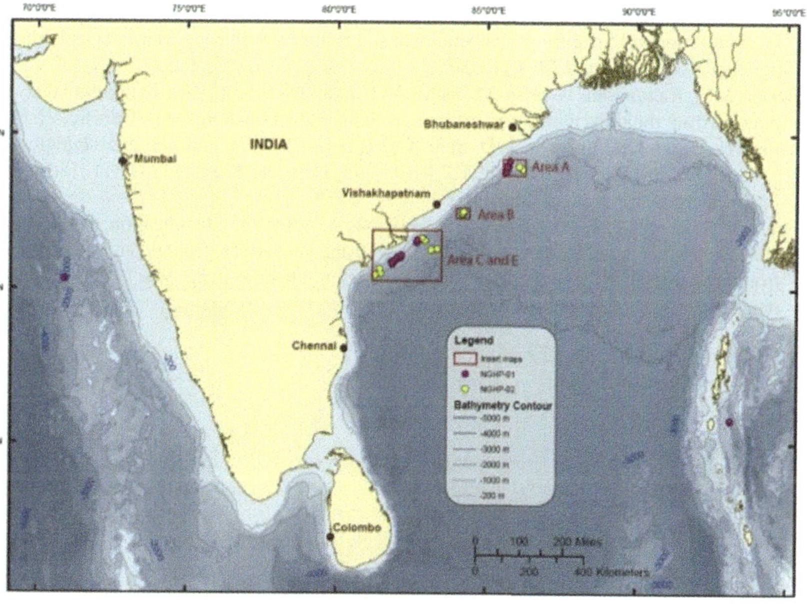

Fig. 14.2 Image depicting logging during drilling of (LWD) sites at NGHP-02 (Jang et al. 2018)

employed to analyze the total costing of the gas production from these gas hydrate fields. The eastern offshore are methane hydrates and are rich source of pure methane. The Indian Government and Petroleum and Natural Gas Ministry has notified the marketing and pricing freedom of the gas produce from these gas hydrate reserves in deep sea water in March, 2016. The methane price was fixed for the specific period (01-October-2018 to 31-March-2019) and was estimated to be USD 7.67/MMBTU on gross calorific value basis (Deepak et al. 2018).

References

Banerjie V, Mittal AK, Agarwal K, Uniyal AK, Chandra K (1994) Carbon isotope geochemistry of petroleum-associated gases in Krishna-Godavari Basin, India. Org Geochem 21(3–4):373–382

Collett TS, Riedel M, Cochran JR, Boswell R, Kumar P, Sathe AV (2008) Indian continental margin gas hydrate prospects: results of the Indian National Gas Hydrate Program (NGHP) expedition 01. In: Proceedings of the 6th international conference on gas hydrates

Deepak M, Kumar P, Singh K, Yadav US (2018) Techno-economic forecasting of a hypothetical gas hydrate field in the offshore of India. Mar Pet Geol. https://doi.org/10.1016/j.marpetgeo.2018.11.016

Dewangan P, Sriram G, Ramprasad T, Ramana MV, Jaiswal P (2011) Fault system and thermal regime in the vicinity of site NGHP-01-10, Krishna-Godavari basin, Bay of Bengal. Mar Pet Geol 28(10):1899–1914

Jang J, Waite WF, Stern LA, Collett TS, Kumar P (2018) Physical property characteristics of gas hydrate-bearing reservoir and associated seal sediments collected during NGHP-02 in the Krishna-Godavari Basin, in the offshore of India. Mar Pet Geol. https://doi.org/10.1016/j.marpetgeo.2018.09.027

Kumar P, Collett TS, Shukla KM, Yadav M, Lall V, Vishwanath K (2018) India National Gas Hydrate Program Expedition-02: operational and technical summary. Mar Pet Geol. https://doi.org/10.1016/j.marpetgeo.2018.11.021

Ramana MV, Ramprasad T, Desa M, Sathe AV, Sethi AK (2006) Gas hydrate-related proxies inferred from multidisciplinary investigations in the Indian offshore areas. Curr Sci 91:183–189

Shankar U, Gupta DK, Bhowmick D, Sain K (2013) Gas hydrate and free gas saturations using rock physics modelling at site NGHP-01-05 and 07 in the Krishna-Godavari Basin, eastern Indian margin. J Petrol Sci Eng 106:62–70

Shukla KM, Kumar P, Yadav US (2019) Gas hydrate reservoir identification, delineation, and characterization in the Krishna-Godavari Basin using subsurface geologic and geophysical data from the national gas hydrate program 02 expedition, offshore India. Mar Pet Geol. https://doi.org/10.1016/j.marpetgeo.2019.05.023

Part III
Gas Hydrates Conclusion

Hydrates Our Hope to Meet Energy Demands

The hydrate crystals have an amazing innate property that these can exist in all three forms of matter depending upon their pressure and temperature condition suitable for their existence. Hydrates project themselves as a brilliant source of minerals available in nature and their property majorly affect the processes important for the existence of the earth. Though the hydrates were discovered approximately 200 years back, but still a very little is known about them, their properties, their capacities of the reserves.

Figure 3 from Chap. 1 presents the natural hydrate discovered from the sea during drilling along the major hydrates reserves identified around the world. The hydrates their structure and composition has been briefly discussed to analyze the depth of hydrates as an energy source. The condition and the properties of hydrates during their formation are the major goals for the scientist and the studies includes:

- To identified the source for the hydrate formation.
- Effective method to control the formation of hydrates in pipelines.
- Technological advancements for better study of gas hydrates as an energy reserve.
- To analyze the effects of hydrate formation on climate, ecology and global change.

It has been discovered from various scientific bodies throughout the world that gas hydrate has great potential in serving as a global reserve of energy. Though in the present scenario there is no technological advancement available to exactly identify and estimate these hydrate reserves. After 200 years of hydrate discoveries the humans have started utilizing these hydrates in various forms like:

- Storage of gases in gas hydrate clathrate
- Transportation of gases
- Desalinization of aqueous solutions

- Hydrate plug in well bore
- Concentrating heavy water
- Carbon dioxide sequestration.

Hydrates are very much beneficial as energy sources but still have some demerits that cannot be looked down and make the use of these gas hydrate system limited. The formation of hydrates plugs in wells and pipelines is the major problem existing and the only solution which is economically feasible is to slow the rate of hydrate formation in wells and pipes rather than expanding money and energy in removing it.